Orbital/Fractional Orbit Bombardment System

The Soviet Globalnaya Raketa

HUGH HARKINS

ISBN: 1541197925
ISBN-13: 978-1541197923

Orbital/Fractional Orbit Bombardment System

The Soviet Globalnaya Raketa

Createspace Independent Publishing Platform
United States

ISBN 10: 1541197925
ISBN 13: 978-1541197923

This volume first published in 2017

The publisher and author would like to thank all organisations and services for their assistance and contributions in the preparation of this volume: Yuzhnoye State Design Office; S.P. Korolev Rocket and Space Corporation Energia; JSC MIC Mashinostryenia (Joint Stock Company Military Industrial Corporation Scientific and Production Machine Building Association); JSC NPO Energomash; OSC KBKhA (Konstruktorskoe Buro Khimavtomatiky, Krunichev State Research and Production Space Centre), National Space Agency of Ukraine, State Space Corporation ROSCOSMOS, National Aeronautics and Space Administration, NASA, United States Department of Defence, Central Intelligence Agency, and the Ministry of Defense of the Russian Federation

CONTENTS

INTRODUCTION

This volume sets out to detail, from the historical and technological perspectives, the fully developed and deployed - 1969-1983 - Soviet 8K69 fractional orbit bombardment system/depressed trajectory intercontinental ballistic missile which will entail a survey of the 8K67 family of intercontinental ballistic missiles from which the 8K69 was developed. The volume also details the rival UR-200, UR-500 and GR-1 orbital missiles complex developments that preceded the 8K69 orbital weapon.

The road, factual and propaganda, that led to the development of the 8K69 orbital weapon complex is detailed as is the missile complex itself, along with the flight test and development program leading to actual deployment of the system. The volume also briefly covers the non-realized concept of the multiple orbital bombardment system and the byproduct of the 8K69 orbital weapon complex that was the Cyclone-2 and Cyclone-3 space launch vehicles.

It should be noted that at varying times throughout the text the weapon systems will be referred to under their Soviet service and or manufacturer designations as well as, at appropriate times, under their NATO designations, the latter of course being accepted by the Soviet Union for use in arms limitations and other treaties.

All technical data concerning the respective weapon systems and their components have been provided by the respective design bureau/offices, as has much of the imagery and graphics with additional impute from United States intelligence agencies and defense department, the space agencies of the Ukraine and the Russian Federation, United States National Aeronautics and Space Administration and the Ministry of Defense of the Russian Federation.

1

THE BIRTH OF THE ICBM TO THE ORBITAL BOMBARDMENT SYSTEM – MOBS & FOBS

The Soviet 8K67 ICBM (Intercontinental Ballistic Missile) and the 8K69 FOBS (Fractional Orbit Bombardment System), together with the later 8K67P ICBM, constituted the R-36 missile complex family. This family of Soviet strategic missiles was developed in response to the development of the Titan II silo based ICBM in the United States in the early 1960's, there being no immediately available Soviet equivalent to this super-heavy American missile that was capable of firing a heavy nuclear payload over vast distances. An added impetus, in regards to the 8K69, was provided by the inordinate advantage NATO (North Atlantic Treaty Organization) had in deployment of IRBM (Intermediate Range Ballistic Missiles) near to the borders of the USSR, something the Soviets, by virtue of geography, were not able to do in regards to the CONUS (Continental United States). This put the Soviet Union at the disadvantage of having to counter ballistic missile attack from not only the traditional northerly, over the pole, direction of CONUS based ICBM's, but also, in regards to attack from Eurasian NATO territory, westerly and southerly directions.

Growing up in the late 1970's and early 1980's, endowed with the naivety of a young boy who viewed the prospect of global nuclear conflict as something exciting rather than being the holocaust it would be perceived as with the onset of manhood, it is easy to look back on those late period Cold War years with a certain nostalgia. Discussions with ones friends, the fantasy finger on the nuclear button and the two minute warning, neither of which had any real basis in fact, now seem a rather ludicrous waste of time. In much the same way, government pamphlets, delivered to every residence, explaining how to survive a nuclear attack, were, on looking back, a waste of good paper. While it would have been logical, with hindsight, to imagine similar conversations and pamphlet reading taking place in the Soviet Union and Warsaw Pact, such things did not occur to young minds conditioned to the perceived reality that they (the Soviets) were waiting for that opportune moment to embark upon a decades old plan (no such plan existed, the Soviet policy being defensive against NATO) to destroy what was then, perhaps, spuriously termed the free world.

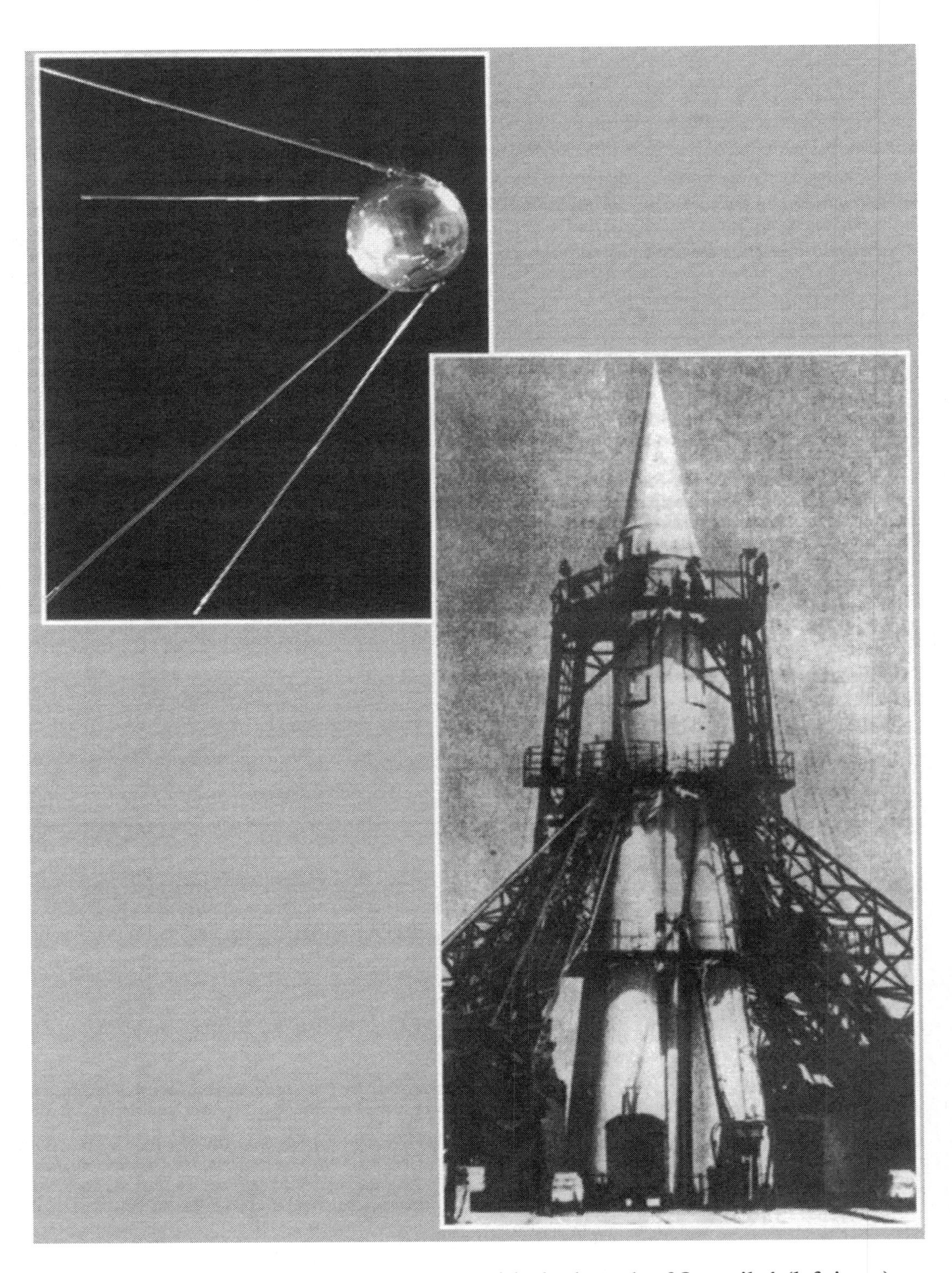

The dawn of the age of space travel came with the launch of Sputnik 1 (left inset) on a converted R-7 ICBM on 4 October 1957. This would lead to a plethora of Earth orbit, interplanetary and Moon craft, as well as the fractional orbit/orbital weapon system of the late 1960's. Energia

The seeds of the climate of fear prevalent in those late Cold War years were sown two decades before when the Soviet Union and its NATO rivals were vying for ballistic missile supremacy with the added prize of access to space. This atmosphere of mutual mistrust and suspicion was prevalent within the populations of the Eastern and Western blocks aware that war between the respective power blocks (the Soviet dominated Warsaw Pact and United States dominated NATO) could lead to Armageddon such as human kind had never before witnessed.

While this volume is not intended to serve as a history of Soviet rocket development leading to the 8K69 FOBS, an overview of such development milestones would be pertinent. It is an indisputable historical fact that the post war rocket programs in the Soviet Union and the United States that would lead to the dawn of the ICBM and space launch vehicles were born out of the ashes of World War II Germany and in particular the A-4 (V-2) rocket, which was tested and further adapted by the Soviets and Americans. In the Soviet Union this lead to a 14 April 1948 Soviet Government Decree authorizing development of what would become the R-1, the first Soviet ballistic missile derived from the A-4 (FAU-2).

The development of the R-1 was enabled by the formation of several test facilities and manufacturing centers as had been decreed by the Soviet government on 13 May 1946, as part of the overall effort to develop a LRBM (Long Range Ballistic Missile) under the direction of Chief Designer S.P. Korolev. A decree of the Soviet minister of armaments, D.F. Ustinov, issued on 26 August 1946, authorized the formation of an organizational structure under NII-88, effectively putting in place Department 3 of the Special Design Bureau (now the S.P. Korolev Rocket and Space Corporation Energia) under the leadership of Korolev. Korolev would become the architect of not only the Soviet Union's early long range ballistic missile programs but also that nations space launch vehicle programs that signaled the dawn of the space age.

The early beginnings with the A-4, the first Soviet launch of which, a missile assembled from various units of captured German rockets, occurred on 18 October 1947. Soviet flight testing of the A-4 was concluded in late 1947 and the first R-1 launch was conducted on 17 September 1948. This launch was considered a failure, as although it almost achieved the specified design range a malfunction of the off-nominal function of the control system sent the missile off course. The first completely successful launch and flight of the R-1 was conducted on 10 October 1948 and the first phase of R-1 flight testing was completed later that year paving the way for the missile system to enter operational service in 1950. The R-1, with a maximum firing range of 270 km, would later fall into the group of missiles termed SRBM (Short-Range Ballistic Missiles) which covered vehicles with a maximum firing range up to 1109 km (599 nm).

In a prelude to space flight testing an R-1 derivative, the R-1A, was utilized from 21 April 1949 to launch a series of six upper atmosphere missions (altitudes of 100 km) to release instrumented containers that conducted atmospheric tests during their parachute descent to terra firma.

A new ballistic missile, the R-2E (Experimental) was launched on 21 September 1949 to test systems for the projected R-2 missile complex, which had a detachable nose cone warhead housing section; this missile complex entering service in 1951.

The OKB-1 single-stage R-1 was the first operational Soviet ballistic missile. With a firing range of 270 km the R-1 would fall into the category of what became known as short-range ballistic missiles. The conventional warhead mass was 1075 kg. Energia

Soviet missile developments of the late 1940's and early 1950's would lead to a multitude of designs including the R-5, the first of the Soviet Union's nuclear armed ballistic missile complexes. Design of the R-5 was completed in October 1951, leading to a maiden launch on 15 March 1953, design work on the R-5M commencing that same year. Development of the R-5M was authorized by a governmental decree issued on 19 April 1954. This missile complex, the maiden launch of which occurred on 21 January 1955, was capable of carrying a nuclear (special) warhead to a distance of 1200 km bringing the weapon into the MRBM (Medium Range Ballistic Missile) category, which covered such missiles with the firing range coverage of 1111 km (600 nm) to 2778 km (1500 nm).

A number of R-5 derivatives were developed as geophysical launch vehicles, including the R-5A models which were launched in the period 1958-1961 and the R-5B which were launched in the period 1964-1975.

Although R-5 variants had been instrumental in bestowing upon the Soviet Union a nuclear armed ballistic missile capability, it was with the R-7 that the dawn of the ICBM and space travel would become a reality, this missile complex at a stroke catapulting the Soviet Union ahead of its competitors in both fields. The large R-7 missile complex had been developed under a Soviet government decree issued on 13 February 1953. This had called for development of what would become a two-stage missile of some 170 t mass, incorporating a separating nose section containing a nuclear warhead of about 3000 kg mass (the warhead capability specification was changed in October 1953 to incorporate a fire charge mass of up to 3000 kg with a total warhead mass of up to 5500 kg) capable of striking targets 8000 km distant from the launch site, bringing it well beyond the 5556 km milestone that would define the ICBM category of missiles.

The R-7, which entered service as an ICBM in August 1957, would lead to a whole family of space payload launch vehicles that would include the Vostok and Molniya, which would lead to the Soyuz launch vehicle. The first ever payload delivery to Earth orbit was conducted on 4 November 1957 when a modified R-7 ICBM launched the world's first artificial satellite, Sputnik 1 (PS-1), setting the Soviet Union apart from the rest of the field in regards to development of space launch vehicles, in effect the stone that started the ripple of shock waves that resonated around the western world in the form of the *beep, beep, beep…* transmitted from the satellite back to Earth. The R-7 launch vehicle derivatives would be instrumental in the Soviet Union's capacity to maintain an undisputed lead in unmanned and manned spaceflight for several years. Such launch vehicles proved to be suitable for near Earth orbit unmanned and manned space flights, as well as several types of unmanned Moon missions and interplanetary probes to Venus and Mars.

While OKB-1 had certainly been the most high profile of the Soviet ballistic missile designers, other design bureau were heavily involved in the development of these promising strategic strike platforms. In this regard, the fledgling NII-88 forwarded plans and research data on what was referred to as a 'high boiling' missile on a par with the OKB-1 R-5, to Vasily Budnik, Chief Designer at Dnepropetrovsk Factory No.586. This early beginning would lead to the development and building of the prototype of the R-12 long-range ballistic missile under OKB-586, also referred to as SDB-586, as Factory No.586 had been renamed in 1954 ((now the Ukrainian Yuzhnoye State Design Office), headed by its chief designer, Mikhail Yangel. The first R-12 launch was conducted on 22 June 1957. This missile complex, with a maximum firing range of 2080 km, fell into the MRBM category.

The R-5 MRBM (R-5M illustrated) was the first Soviet ballistic missile complex to be armed with a nuclear warhead. Energia

The R-12 MRBM (top) and R-14 IRBM (above) were among the first generation ballistic missiles developed by OKB-586. Yuzhnoye State Design Office

The increasing capabilities of ballistic missiles, particularly in warhead mass and range, would be instrumental in the move away from plans to build large fleets of supersonic intercontinental bombers in the mold of the Myasishchev M-50 'Bounder'.

Continued research and development would lead to the R-14 which was endowed with a maximum firing range of 4500 km, more than twice as much as that of the R-12, bringing this missile into the IRBM category that covered ranges from 2778 km up to 5556 km. This missile complex, the first launch of which was conducted in June 1960, would introduce the UDMH (Unsymmetrical Dimethylhydrazine) propellant, which had a desirable 'quality, spontaneous ignition (hyperbolicity) of the components of the propellant mixture. By contrast, first generation ICBM's like the R-7, burned the old mix of kerosene and liquid-oxygen. Accuracy of the R-14 was increased over its forebears and the probability of instrumentation errors was reduced by the incorporation of a gyro-stabilized platform within the autonomous inertial control system.

Further research and development led to the R-16, which, with a maximum firing range of 13000 km, was OKB-586's first ICMB to reach production. The first launch of this missile complex was prepared for under an atmosphere of accelerated pace. During these preparations a faulty control system cable network was the catalyst for a premature start of the missile second stage sustainer resulting in the missile being destroyed in a catastrophic explosion on the launch pad. Following redesign to eliminate the faults an R-16 missile was successfully launched in February 1961.

While the R-12, R-14 and R-16 constituted the first generation of Soviet ballistic missiles developed by OKB-586 the bureaus next steps on the development ladder would lead to the second generation R-36 missile complex and the world's first deployable orbital/fractional orbit weapon system.

Having found itself at a major strategic disadvantage in terms of strategic strike capability during the Cuban Missile Crisis of October 1962, the Soviet leadership determined that such a situation was untenable leading to a strategic armaments program of unprecedented scale in which the Soviet Union intended to not only achieve parity with the United States, but overtake that nation to become the world's premier nuclear power in terms of both delivery vehicle numbers and throw weight. To this end the Soviet ICBM force levels, along with other elements of its strategic strike forces, not least of which was a massive SSBN (Nuclear Powered Ballistic Missile Submarine) build program, increased through the late 1960's and through the 1970's. The latter decade would see the Soviet Union overtake the United States (the US ICBM force was predominantly focused on the lightweight silo based Minuteman variants with smaller numbers of Titan heavy ICBM's) in terms of numbers of deployed ballistic missile launchers available. The United States still had an advantage in its ability to target Soviet territory with MRBM and IRBM's whereas the Soviets could not reciprocate in regards to the CONUS. The Soviet's however did possess a large force of MRBM and IRBM's, the latter estimated at around 700 by the late 1960's, in which they could deploy extensively against targets on the Eurasian (includes the United Kingdom) land mass in the event of conflict with NATO, and such missiles could be targeted against other potential adversaries such as China.

While ICBM was the buzzword, more accurately the buzz acronym, of nuclear Armageddon discussions, reasonable, if, in regards to Soviet systems, not always accurate, data being available, there was of course no meaningful knowledge of such systems as orbital weapons in the public domain during the late period Cold War years. Such weapons systems were, it being the public perception, the object of science fiction, forming the story centerpiece of then modern culture science fiction movies such as 'Meteor', 1979.

Even prior to the dawn of the space age that began with the successful launch of the Sputnik 1 artificial Earth satellite onboard the R-7 ICBM derivative launch vehicle on 4 October 1957, there was detailed discussion of the use 'outer space' as was then the term for what lay beyond the Earth's atmosphere, to base orbiting platforms capable of launching atomic bombs on unsuspecting targets back on Earth. Among the most prominent of such early theories to come out of the west were those touted by the former German scientist Dornberger, who, with the end of World War II in Europe in summer 1945, was transferred to the United States to work on American funded rocket programs. This particular scientist was scoffed at as a former Hitlerite by the Soviet Union which held the view that he was nothing less than a deranged Hitlerite madman who, now under his new masters, continued to envision the same destruction of Bolshevism advocated by the former German Chancellor, Adolf Hitler, universally considered responsible for the mass murder of tens of millions of people, during the terrible war years collectively referred to as World War II, but known as the Great Patriotic War in the Soviet Union.

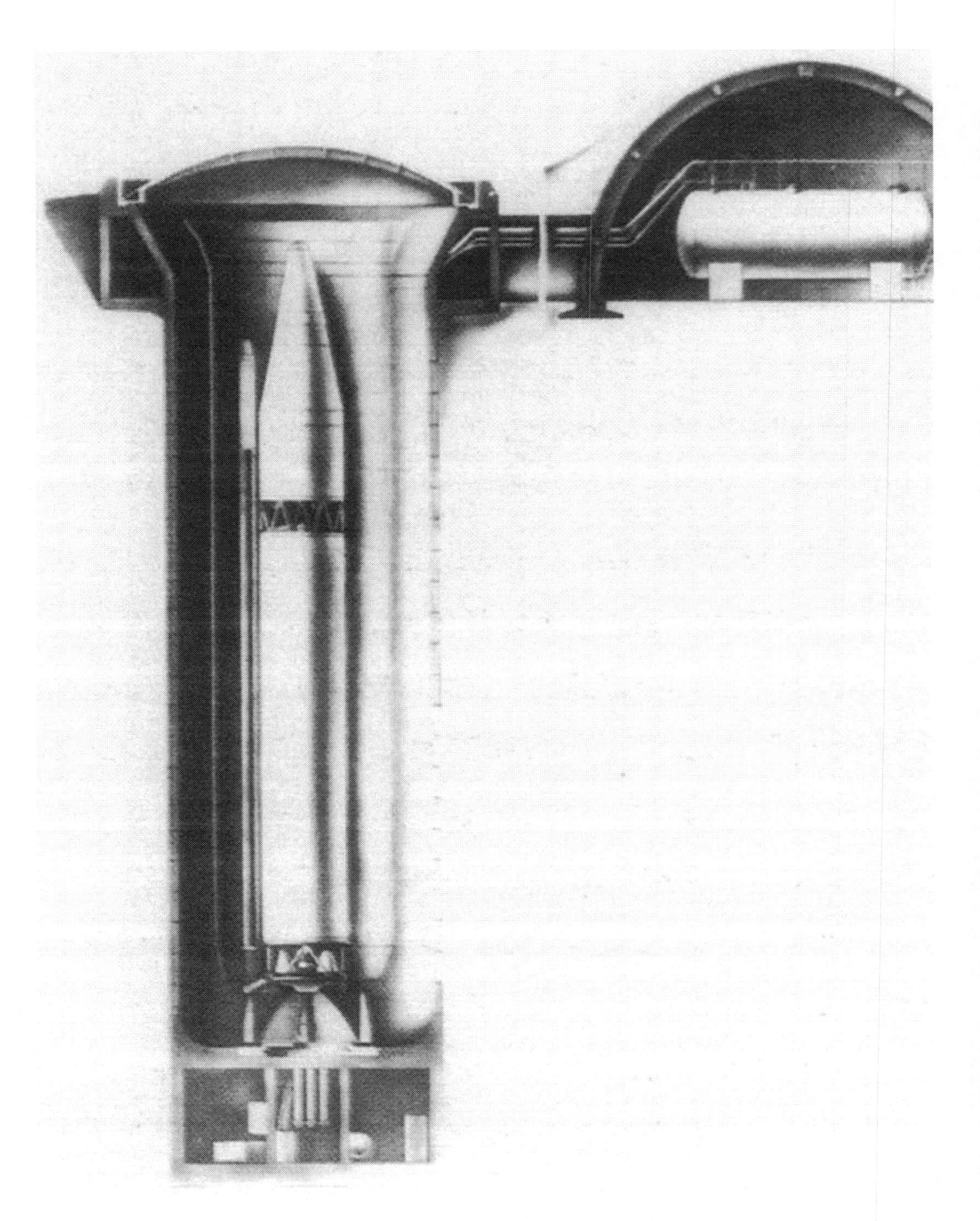

The R-9 was a first generation silo based and launched ICBM that was of insufficient mass to be considered as a viable platform for development as a FOBS or indeed a MOBS. Energia.

The R-16 first generation ICBM had a firing range of 13000 km, the first successful launch being conducted in February 1961. Yuzhnoye State Design Office

The fact that Hitler's victims predominantly came from Eastern Europe and the USSR was not lost on a Soviet Union perplexed at such statements coming from the likes of Dornberger whom it appeared clear by virtue of those same statements was still envisioning the destruction of the Soviet Union. The United States intelligence document on 'Soviet Orbital Rockets', 62-104279, stated that as early as 1948, almost a decade before the first successful launch of an artificial satellite, Dornberger had proposed that, in the event of developments allowing access to space, "the earth be surrounded by hundreds of artificial satellites in the form of nuclear bombs." The document then went on to directly quote, supposedly Soviet writings of a Dornberger quote, that "the orbits of these bombs must be laid over Russia. I see no reason to prevent us [presumably meaning the United States] from doing so... In case of war we will not have to change orbits. I think that it would be possible to launch these bombs with sufficient precision." Perhaps most troubling from this text from a Soviet point of view was that in 1948 the Soviet Union was not yet a nuclear power (the first Soviet atomic device was detonated in 1949) and here was what appeared to be detailed sinister plans for the destruction of the country's infrastructure in the event of the outbreak of even a local conventional war.

It was such visions of the use of space as a platform for the launching of atomic weapons to destroy the cities of the Soviet Union that spurred Soviet attitudes towards achieving superiority in space in both the fields of exploration and, if need be, for military advantage. For this the Soviets would capitalize on their superiority in rocket development evinced by the launch of Sputnik 1, which was followed by what seemed to be Soviet success after Soviet success in the areas of unmanned and manned spaceflight over the next half decade or so, encompassing a period known as the early space age.

During the period of the early space age there were many idle references in the Soviet and western press about the feasibility of space based weapons, such references to satellites that could launch hydrogen bombs at targets on Earth continuing through the early 1960's, although there was no official references to such theorized weapon systems. Soviet research efforts were being invigorated by intelligence evidence coming out of the United States on such space based weapons programs as Project 'Saint', Project Sped' and Project Bambi, not to mention the potential usage of the X-20 Dynasoar spaceplane, then under development, as what was being unofficially referred to as an orbital bomber.

Concern about the possible development and deployment of orbital weapons by the United States led to Soviet overtures that proposed a ban on the deployment of offensive weapons in space. In reference to possible future weapons applications in space, Professor A. Rybkin, in an article published in the Soviet publication Red Star on 22 March 1958, certainly acknowledged that it was feasible to place "atomic and hydrogen" weapons, presumably bombs, on orbiting artificial Earth satellites. Rybkin's statement went on to read "As the foreign press has stated, satellites can carry atomic and hydrogen weapons. Considering the difficulties in intercepting and destroying a satellite in space, it must be kept in mind that it could be a dangerous weapon in the hands of the aggressor. The Soviet Government's proposal which poses banning the utilization of space for military purposes also includes conditions for international control which would preclude the possibility of artificial earth satellites being used for aggressive purposes."

There were of course many references in the western press in regards to the development of orbital weapon systems of the type that would deploy atomic and hydrogen bombs. Several such references were picked up on by a Red Star article by Soviet military theorists Zheltikov and Larinov, published on 29 November 1961. One extract quoted in the intelligence document on 'Soviet Orbital Rockets', 62-104279, read, "in the very near future - if not already - mighty hydrogen bombs will almost certainly be installed in earth satellites. As a result, space ships or stations will be able to launch rockets against objects on earth".

The first official references to orbital bombardment systems surfaced at a Soviet Trade Unions Congress speech in Moscow by Soviet Premier Nikita Khrushchev on 8 December 1961. Khrushchev dropped passive hints of the feasibility of the Soviet Union developing such systems on the back of recent Soviet space exploration exploits, "if we could send up Yuriy Gargarin and German Titov, we could, of course, replace Yuriy Gargarin and German Titov by other freight and land where we would like to land", stated Krushchev without actually using the term warhead.

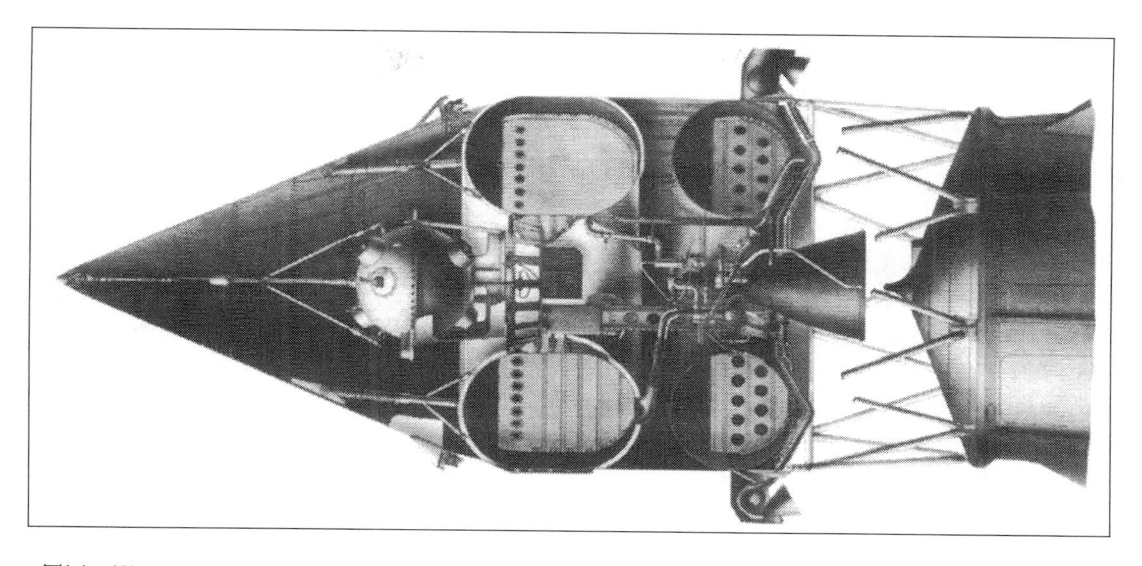

This illustration shows an unmanned lunar station atop a Block E booster. In the early 1960's it seemed logical that such payloads could be replaced by a nuclear warhead armed Earth orbiting station that would be placed into a low-earth orbit for several months at a time. Energia

Another version of Khrushchev's speech, which, according to intelligence document on 'Soviet Orbital Rockets', 62-104279, was carried in TASS, apparently stated, "A devastating retaliatory blow will be dealt against any aggressor, because the means of delivering thermonuclear weapons are now so perfect that they can be delivered to any point on the globe." The statement continued to include the quote mentioned above regarding replacing cosmonaut's with other payloads.

It was Khrushchev who would publically introduce the phrase 'global rocket' (globalnaya raketa) which he used during a campaign for election speech on 22 March 1962. The assertions of unlimited range provided few indication of the characteristics of such a weapon complex, "the new global rocket can fly around the world in any direction and deal a blow at any set target", stated Khrushchev, who continued by referring to its "invulnerability… practical invulnerability…" to antimissile defenses in reference to a then proposed US ABM (Anti-Ballistic Missile) system.

Although the term 'global rocket', probably influenced by OKB-1's planned GR-1 rocket, was, in the early 1960's, considered a generalization, it could, perhaps, be seen as a reference to the type of weapon system that would, several years later, be defined as a FOBS rather than what was considered at the time, that such references, along with the term orbital weapon, referred to spaceships of the time armed with hydrogen bomb warheads that would remain in orbit for indefinite periods until such time as they were called upon to be launched against pre-selected targets.

A Soviet statement on the peaceful use of space that was printed in a volume entitled 'Military Strategy', edited by Marshal Sokolovskiy and signed to the press on 24 May 1962, included the following text reproduced in intelligence document on 'Soviet Orbital Rockets', 62-104279, "Our achievements in space research serve peace, scientific progress, and the benefit of all mankind on our planet". The text continued, "Soviet space flights are an expression of the unstinting efforts of the entire Soviet people to achieve lasting peace on earth. However, the Soviet Union cannot disregard the fact that American imperialists subordinate space research to military purposes and that they plan to use space to accomplish their aggressive purpose – a surprise nuclear attack on the Soviet Union and the other socialist countries. Consequently, Soviet military strategy acknowledges the need to study the use of space and space vehicles to reinforce the defense of the Socialist countries." Such statements may serve to show that the paranoia that had gripped America in regards to communist plots to take-over the country, the bomber gap and the missile gap, were also present in the Soviet psych in the first half of the period defined as the Cold War.

In the four months or so following the Cuban Missile Crisis, which bestowed upon a worldwide public an illusion of a Soviet back down in the face of nuclear war with NATO (in fact the opposite could now be considered true as the Soviet removal of its non-operational R-12 IRBM from Cuba came at the price of a secret US agreement to remove its Jupiter IRBM's from Turkey which shared a border with the Soviet Union, in effect a move in the strategic balance towards the Soviet Union), a number of statements came out of the Soviet Union that hinted that orbital weapons were at the very least under development if not already deployed. Prominent among such was the wording of an article by Major General G. Shatunov published on 27 January 1963, in which the implication was that, at that time, the Soviet Union had the capability to deploy rockets on artificial Earth satellites and launch same, presumably towards targets on Earth. The article stated, "It has now become possible to launch a rocket from an artificial earth satellite on command from earth at any time and at any point in the satellite's trajectory". There were of course other references, verbal and in print, about the feasibility of orbital weapon systems in both the Soviet block and within NATO member nations.

Published in Izvestiya on 17 November 1963, Soviet Marshal Krylov stated of the USSR's strategic rocket forces, without referencing 'global rockets', that it was a force with high accuracy and the ability to fly flight "trajectories within parameters which make nuclear rocket attack both unexpected and inevitable". This was, perhaps, the first suggestion of a depressed trajectory capability for existing or projected ICBM's, a term that would come to be known as DICBM (Depressed trajectory Intercontinental Ballistic Missile). This suggested to the intelligence agencies a possible move away from orbital weapons to a new type of capability for the ICBM force.

Over the next few years evidence of a deployed orbital weapon was simply not there, leading to intelligence analysts concluding that such projects, if indeed they existed, had not been flight tested or deployed. Such analysis was reinforced by a Red Star article of 27 January 1965 by Colonel. C. Glaznov, which refuted any previous intelligence agency inference that orbital weapon systems were already operational. The article noted that space weapons existed only as projects in the Soviet Union and that was his inference of the United States space weapon programs, of which he stated there "untold numbers".

The first claim by a high ranking Soviet official about the actual possession of an orbital weapon in any form came from Brezhnev who, in a speech to military graduates on 3 July 1965, indicated that the Soviet Union was in possession of a force of not only intercontinental ballistic missiles, but "orbital rockets", that would enable that state to destroy any aggressor nation.

The Soviet leadership utilized both of the non-flight rated OKB-1 designed GR-1 'Global Rockets' in the 1965 May Day parade as part of its misinformation campaign designed to throw confusion among the western intelligence agencies. The GR-1 missile complex had in fact been cancelled the previous year as the focus turned on the second generation OKB-586 8K69 development of the 8K67 ICBM. US DoD

References to 'orbital rockets' had been unofficially made at the May Day parade of 9 May 1965, followed by the following text directly quoted from the publication Wings of the Motherland No.7, dated 20 June 1965, reproduced in intelligence document on 'Soviet Orbital Rockets', 62-104279, "Gigantic orbital rockets-related to the rocket carriers which confidently put into space the space ships in the series Vostok and Voskhod. For these rockets there is no range limit, and the possible might of their nuclear warheads is fantastic." Such statements, which were of course unofficial, hinted that the so called 'orbital rockets' were based on a development of the R-7 (SS-6) ICBM. However, adding to the confusion in the west was the fact that such references were being associated with footage at the 1965 May Day parade of the OKB-1 GR-1 (this missile complex had been cancelled the previous year and was displayed at the 1965 May Day parade for the dual purpose of propaganda and to confuse western intelligence analysts) which would become known in the west under its NATO reporting designation and name SS-10 'Scrag' (SS-X-10). The May Day parade commentary suggested that the orbital rockets, which were related to the rocket that launched the Voskhod II capsule into orbit, could target any area on Earth from all directions and was to all intents and purposes "invulnerable" to antimissile defenses as then forecasted, this obviously being the source for the Wings of the Motherland assertions of same the following month.

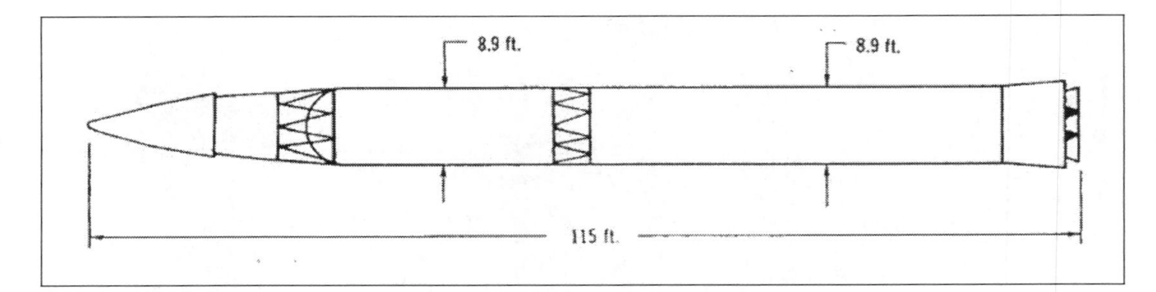

Top: Following the OKB-1 GR-1 debut at the 1965 May Day parade, United States intelligence analysts quickly drew up their plans of the missile complex layout including this relatively crude technical layout with inaccurately estimated measurements of the height and the diameter of the first and second stage boosters. US Gov.

Above: One of the GR-1 non-flight rated missiles is transported past assembled crowds during the May Day parade in Moscow on 9 May 1965. US DoD

The still less than detailed definition of what was meant by 'orbital rocket' was slightly elaborated upon by a Radio Moscow English language broadcast, which was directed at a South Asia audience three days after Brezhnev's speech on 3 July 1965. The broadcast commentator, whilst citing Brezhnev's 3 July speech, stated, "These rockets are shot into a terrestrial orbit from where they are capable of hitting any target on earth when needed. Distinct from other types of rockets, they have practically no flight limit and are capable of carrying super-powerful nuclear charges." Although not an official statement from a Soviet government or military official, this broadcast was, perhaps, the first, albeit circumstantial, evidence that what the Soviet Union was developing, or as some thought, had deployed, was a weapon system that would come to be defined as a FOBS as opposed to a MOBS.

The many references to orbital weapons programs in the Soviet Union, particularly in the early 1960's, was fueled by a deep fear that the United States intended to develop and deploy such systems in pursuance of a plan to dominate space for its own military ends. Already at a strategic strike (nuclear armed long-range bomber and ballistic missile forces) disadvantage, the Soviet Union, with its lead in space deployment capacities that existed in the early 1960's, was determined not to fall into a position of inferiority in regards to the militarization of space. To this end the Soviet Union continued to propose a treaty banning the deployment of weapon systems in space and embarked upon a psychological campaign in an attempt to force the western powers, in particular the United States, to sign up to such a treaty which finally resulted in the Outer Space Treaty, banning the deployment, but not development or testing of such weapons systems as a MOBS.

In this regard it is clear that much of the Soviet rhetoric about the development and or possession of 'orbital rockets' from the late 1950's and early 1960's was aimed at exerting pressure on the United States to sign the proposed agreement prohibiting the orbiting of weapons in space, which, along with the agreement it secured on a partial nuclear test ban treaty, the Soviet's would attribute to contributing significantly to a period of relaxed East West relations in the late 1960's and 1970's.

In evaluating the reasons behind much of the Soviet rhetoric regarding orbital weapons, MOBS can be considered a political tool in that there were probably never any serious plans to develop and deploy such a system, as defined by its Cold War definition, by either of the two power blocks of East and West. Among the reasons for this was the fact that such weapon systems would, considering technology levels of the 1960's and even into the 1970's, have been less accurate and probably have carried lower yield warheads than terrestrial based ICBM's. In addition, they would certainly have had shorter lifespans and higher overall costs than ICBM's and arguably, considering the impetus that would have then existed to develop countermeasures to such weapons, been perhaps equally as vulnerable as terrestrial based ICBM's. Another major factor weighing against the MOBS concept would have been the requirement for sophisticated command and control communications links with terrestrial control centers in much the same was as an orbital space exploration vehicle would require such systems. This increased complexity would be vulnerable to interference and the potential of loss of communications links and ergo control of the system. Such drawbacks and risks increased the weight of argument against development of such weapons, even more so than the prohibitive cost, as the potential problems, political and military, that would have emerged in the event of a loss of command and control with a MOBS warhead carrier would have been prohibitive.

The development of MOBS was not pursued through to significant material work being conducted by either the Soviet Union or the United States as each intended to enter into the treaty banning the deployment of such systems. Although the resultant Outer Space Treaty, which covered not only Earth orbit, but the Moon and all other celestial bodies, was not signed by the various parties in Washington, London and Moscow until 27 January 1967, the treaty entering into force on 10 October that year, there were a number of drafts that were refined from 1966.

Such weapon systems as MOBS would live on only in fiction, forming the story centerpiece of science fiction movies such as the aforementioned 'Meteor', 1979 and more recently with the post-Cold War, 'Space Cowboys' 2000, both of which included storylines that revolved around orbital weapons complexes that would have constituted MOBS under the Cold War terminology, albeit, such an operational weapon system would not have consisted of full scale ballistic missiles as appeared to be portrayed in the movies.

Of course, during the late Cold War period and in the decades after the fall of the Soviet Empire that led to the ending of that Cold War stand-off between East and West, information began to trickle out of the former Soviet Republics of a number of orbital weapons programs, both defensive and offensive. In regards to the latter category only one system attained operational status, the OKB-586 8K69 FOBS, conceived and born out of that multitude of propaganda and here-say about 'orbital rockets' in the early 1960's, which would evolve into the term 'global rockets' that took on more of the characteristics of the system that would be defined as a FOBS. A FOBS would be capable of conducting multiple orbits over the course of several days to as much as several weeks. However, it was distinct from what was regarded as a MOBS in that the latter was classified as a weapon system that would be launched into an Earth parking orbit with one or more warheads on-board which would be targeted and launched form orbit, this requiring a complex command and control system not inherently required for a FOBS system conducting multiple orbits.

2

ORBITAL MISSILE COMPLEX – UR-200, GR-1, UR-500 & 8K69 (R-36 MOD 3)

The 8K67 (NATO reporting designation SS-9-1/2 or alternatively SS-9 Mod 1/2) ICBM (Intercontinental Ballistic Missile) and 8K69 (SS-9-3/SS-9 Mod 3) FOBS/DICBM (Fractional Orbit Bombardment System/Depressed Trajectory Intercontinental Ballistic Missile), together with the later 8K67P (SS-9-4/SS-9 Mod 4) ICBM, collectively constituted the R-36 missile complex family. The R-36 was developed by the Soviet Union in response to the development of the Titan II silo based ICBM in the United States in the early 1960's, there being no immediately available Soviet equivalent to this super-heavy American missile that was capable of firing a heavy nuclear payload over vast distances.

Developed off the back of existing ICBM technology, the 8K69 was the end product of Soviet efforts to develop a FOBS that had transitioned from the theoretical to the practical development stage with the Kremlin's decision to develop such a system from 1961. Initially such a weapon system was to be based on an EDB - Experimental Design Bureau 52 (now JSC MIC Mashinostryenia) design based on the UR-200 launch vehicle. Development of the UR-200 had commenced in 1960 as a first generation space launch vehicle/heavy ICBM, design solutions being worked so that the main engines thrust could be deviated in direction.

During 1961-1964, the OCB(OKB)-154 (now OSC KBKhA (Konstruktorskoe Buro Khimavtomatiky) RD-203 and RD-204 rocket engines were developed for the first stage of the UR-200 along with the second stage RD-205 which incorporated the RD-206 main engines and RD-207 steering engines. All of these rocket engines, which burned storable propellants, introduced the staged combustion cycle schematic which conferred a double combustion chamber pressure of up to 150 kg/cm^2 compared to the 70 kg/cm^2 which was the norm for open cycle engines.

The UR-200 possessed characteristics that appeared to facilitate the potential for development as a so called 'global rocket'/FOBS, and, if the Kremlin deemed such a move necessary, further development as the launch vehicle component of a potential MOBS (Multiple Orbit Bombardment System). Besides the FOBS and MOBS, the

UR-200 was intended to constitute the launch vehicle component for other potential space-based weapons systems such as ASAT (Anti-Satellite) systems. This was an area in which EDB-52 was involved in development work for a potential deployable ASAT interception capability, to which end the bureau designed and built the words first maneuvering satellite, the 'Polyet-1' which was launched in 1963, this being done in conjunction with other development work on ASAT systems.

EDB-52 (later OKB-52), in cooperation with other bureaus including Affiliation 1 of the Machine Building Central Design Bureau (now Salyut Design Bureau, a part of Krunichev State Research and Production Space Centre) developed the UR-200 as an ICBM. The missile was also intended to form the basis of an orbital weapon complex and space payload launch vehicle, but the program was cancelled as the technology level was overtaken by second generation missiles. Krunichev

The first launch of a UR-200, which had a launch weight in the order of 140 tons, was conducted on 4 November 1963, less than two months after the first launch of an 8K67 ICBM of the OKB-586 second-generation R-36 missile complex. This was followed by a further eight successful launches that would validate the successful operation of the rocket system, which, however, fell short on the operational requirements specified for the system, particularly if intended to be employed as a FOBS/MOBS launch system. The major shortfall was that it was not able to stand for extended time periods after it had been fueled. This, together with other pre-launch preparation requirements, including the need to be transported from the storage facility, sealing and the above mentioned fueling, added up to a system very vulnerable to a NATO first strike. The shortfalls in capability would result in the programs termination as an ICBM/space payload launch vehicle and, ultimately as a potential platform for any future orbital/fractional orbit or ASAT weapon complex.

While the UR-200 program was still moving forward, OKB-1 (S.P. Korolev Rocket and Space Corporation Energia) was in the early design stages of its GR-1 program, which was officially authorized by a decree of the Soviet government dated 24 September 1962, on which date OKB-1 formally embarked upon work related to the program, informal work having commenced prior to this.

OKB-1's GR-1 'Global Rocket' appeared more promising than the UR-200 in regards to the intended FOBS mode of operation. Energia

OKB-1 was already heavily committed to development of the R-9 ICBM. This missile complex had a firing range of 12000-13000 km, but was, with its launch mass of 81 tons, an unsuitable launch vehicle for use as a viable FOBS. The R-9, the first successful launch of which was conducted on 21 April 1961, would be fully developed and deployed as an ICBM from 21 July 1965.

In theory the basis of development of the a system under the GR-1 program was to provide a strike capability that had several desired traits including a so called invulnerability to a future ABM (Anti-Ballistic Missile) system that was under development in the United States in the early 1960's. It had been realized that space based strike systems, or a ground based ballistic missile following ballistic paths with altitudes above 1000 km, could be quickly detected as they followed the basic laws of motion that equally made it reasonably straightforward to predict their likely impact area and ergo the intended target. This data would provide a developed and deployed ABM system, including employing countermeasures, a much more realistic chance of at least attempting to combat the incoming warheads.

The GR (Global Rocket) concept would be designed to provide a measure of invulnerability to detection through several means. At the low operating altitudes of 150 km or less, the GR would be more or less immune from detection until about 500-600 km from the impact point whereas a typical ballistic trajectory ICBM could be detected from anywhere in the region of 4000-8000 km from the intended impact point. This effectively translated to a difference in potential engagement time against an incoming warhead of as little as 2 minutes for a GR as opposed to the typical 12-15 minutes that was considered to be available against a typical over the pole ballistic trajectory ICBM flying an intercontinental attack profile.

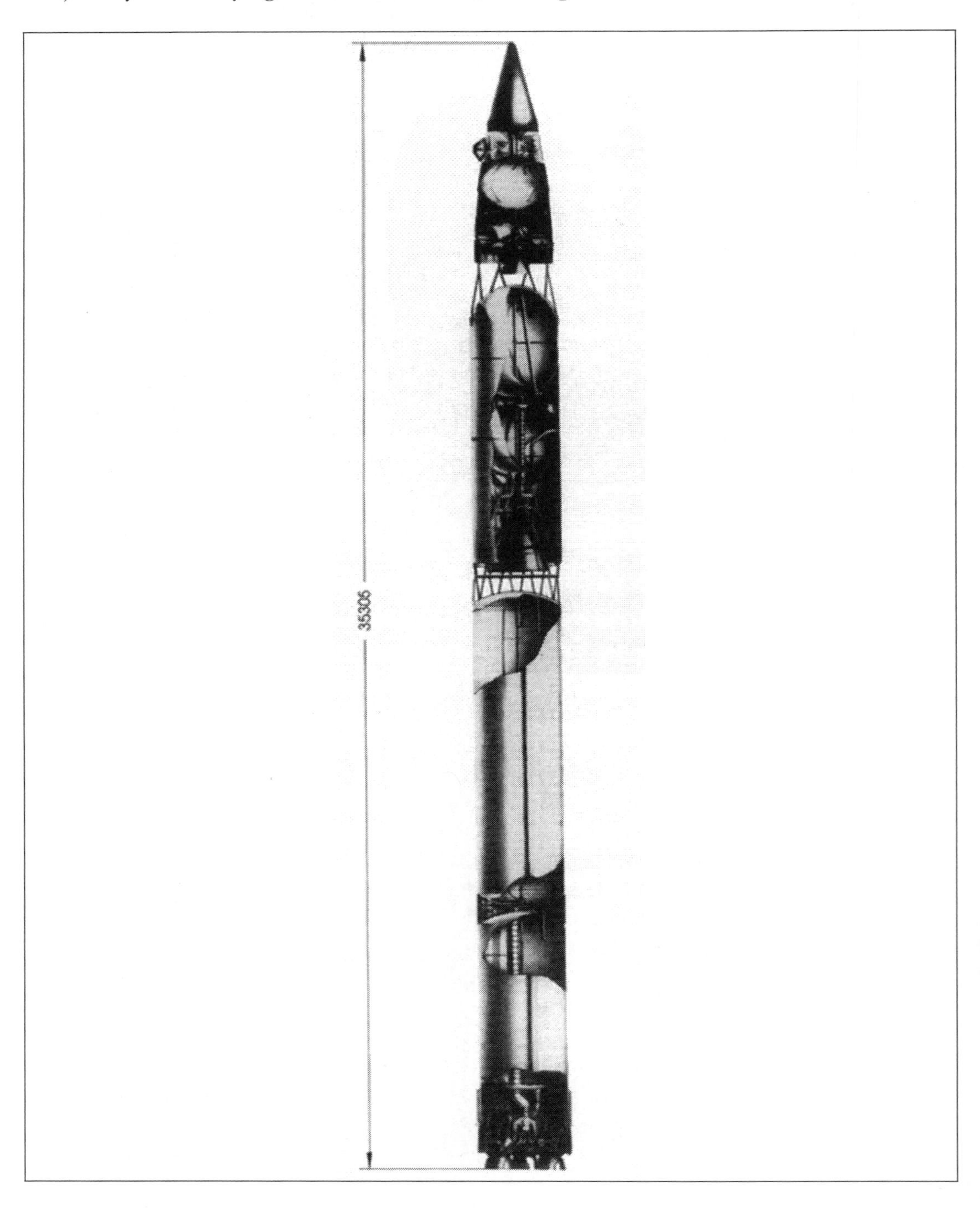

Technical drawing of the OKB-1 GR-1 'Global Rocket' showing an overall height of 35306 mm (35.3 m). Energia

In an atmosphere of Cold War misinformation and misdirection, the GR-1 was presented to the public in the year after its cancellation, although it was portrayed as an operational system to mislead western intelligence agencies. Energia

OKB-1's GR-1 adopted a three-stage configuration and was to be similar in operation to the Energia R-9A ICBM. The starting weight of the GR-1 was 117 tons and 147 tons with traction control on the ground. The missile was designed to inject a payload with a 2.2 MT (Megaton) warhead into low Earth orbit and deliver same to the target after either a fraction of an orbit, or potentially multiple orbits, with an estimated accuracy of ± 5 km in range (front and rear of the target) and ± 3 km on what was termed lateral deflection (either side of the target).

The major fall-off in accuracy inherent in the GR concept could, according to Energia, have been overcome by a system referred to as RDGCH (head movement control part) which was designed to increase aerodynamic drag. This would effectively increase warhead accuracy in range (front and rear of target) if deployed at some point in the flight following braking during the downward trajectory of the reentry, all alterations being calculated by the on-board automatic control system.

The GR-1 propulsion system, based on that developed for the 8K77 missile of the R-9 complex, included stage I engines designed by OKB-456 headed by V.P. Glushko (now NPO Energomash) and stage II engines designed by OKB-154 headed by S.A. Kosberg. For the third-stage, OKB-1 designed the 8D726 engine. Ground tests associated with the 8D726 engine commenced in 1963, a total of either 88 or 230 (two values appear to have been provided in various documentation) such engines being produced and tested in about 500 different test runs. The GR-1 orbital weapon system was designed to be compatible with the starting systems and other ground based systems and infrastructure in place for the R-9A ICBM. However, a purpose designed launch pad was created at Site 51 at Baikonur cosmodrome in Soviet Kazakhstan, in close proximity to the launch pad for the OKB-1 R-7 rocket.

In preparation for planned flight testing of the GR-1 complex an extensive ground test phase ramped up in 1962, this including the building of two full-scale non-flight rated missiles. These would be presented at such public events as the Soviet Red Square May Day parade of 1965, bestowing upon foreign intelligence services the mistaken belief that the system was either in flight test or already deployed. However, far from being deployed, the GR-1 had not attained flight test status. The program encountered a number of not insurmountable technical issues, including with integration of the OKB-726 developed NK-9 engines employed in the missile first stage. However, the major problem, as was the case with the UR-200, was the inability to stand for long periods of time after refueling. For this reason foremost among others the 8K713 missile of the GR-1 system was cancelled in 1964, along with the 8K513 missile intended to be the launch vehicle for the Soviet Unions planned ASAT, neither the 8K713 or 8K513 attaining flight status.

While the UR-200 and GR-1 had both fallen by the wayside, Yangel was pushing on with a development of a potential global rocket modification of the 8K67 missile of the R-36 ICBM complex and other design houses put forward designs. Following the cancellation of the UR-200, development had continued of a larger launch vehicle in support of the Soviet manned space program, for employment in such areas as launch vehicle for Space gliders and rocket gliders. This research led to the initiation of development of the 500 ton class UR-500 launch vehicle designed and developed under V. Chalomey of Affiliation 1 of the Machine Building Central Design Bureau in cooperation with OKB-52. The RD-206 engine of the UR-200 effectively became the prototype of the RD-0208 and RD-0209 engines used in the second stage of the UR-500 which could orbit a payload of up to 20 tons. The development program was aimed at testing two and three stage variants during the period 1965-1968, including the launch of four space stations to conduct research into high and super high energy particles. The first launch, which was conducted on 16 July 1965, placed a Proton research satellite into LEO (Low Earth Orbit).

The Proton-K (UR-500K) launch vehicle was developed from the UR-500 launch vehicle which was initially designed to form the basis of an orbital weapon complex launch vehicle following the demise of the UR-200 and GR-1. Krunichev

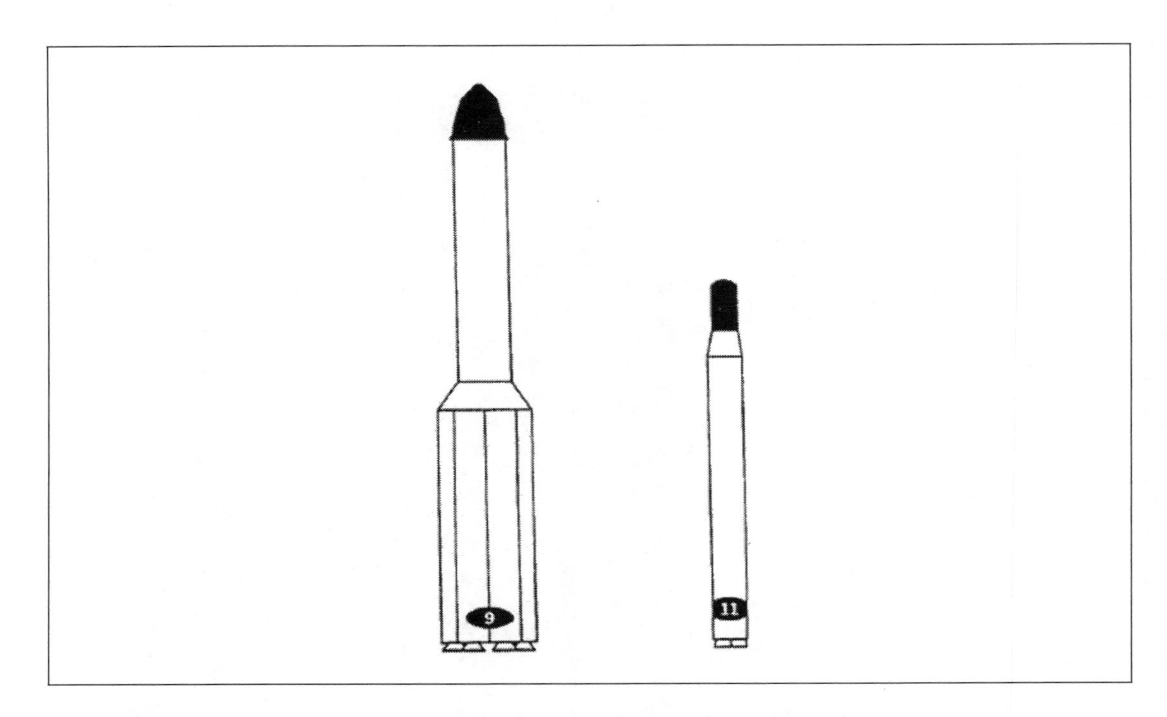

This edited graphic from a US intelligence NIE document shows the relative difference in scale between the Proton launch vehicle (allocated the NATO designation SL-9) that evolved from the UR-500 and the launch vehicle development of the 8K69 (allocated the NATO designation SL-11). US Gov.

Flush with the success of the two-stage UR-500 design, the design team embarked upon further development of the planned three stage variant which was aimed at launching manned circumlunar missions (not the planned manned Lunar landing of the N-1 program) under Project LK-1, this emerging as the UR-500K heavy launch vehicle, which would later carry the name Proton, with a take-off mass of 700 metric tons when fitted with the Block D upper stage. Like the UR-200 before it, the UR-500 launch vehicle had been designed with the potential for development as a strategic ballistic missile complex, and, or, as a 'Global Rocket' armed with what was termed 'global warheads', that would be carried into orbit and be maneuvered in Earth orbit before being deorbited overs a specific target area. However, the UR-500K, the first launch of which occurred on 10 March 1967, was developed purely as a space payload launch vehicle as the Soviet Union had selected the OKB-586 8K69 as the basis of its Global Rocket a few years previous.

Two separate orbital maneuvering warheads were developed for the UR-200/UR-500 orbital weapon programs, the AB-200 for use on the IBM UR-200 and the AB-500 which would arm the larger IBM UR-500 missile complex. Such warheads were designed in such a manner as to be capable of maneuvering in Earth's atmosphere after de-orbit to increase accuracy. To this end OKB-52 developed the MP-1 vehicle which was to be capable of using aerodynamic control to maneuver in Earth atmosphere at high hypersonic speeds, this having been accomplished in what was described as a successful test in 1961 when the UR-200 was under development.

Like the IBM UR-200, the IBM UR-500 complex was never seen through to fruition. The IBM UR-200 program had been considered not technically viable while the larger IBM UR-500 was dropped in favor of the more, near term, technologically attainable and less costly 8K69 of the R-36 missile complex.

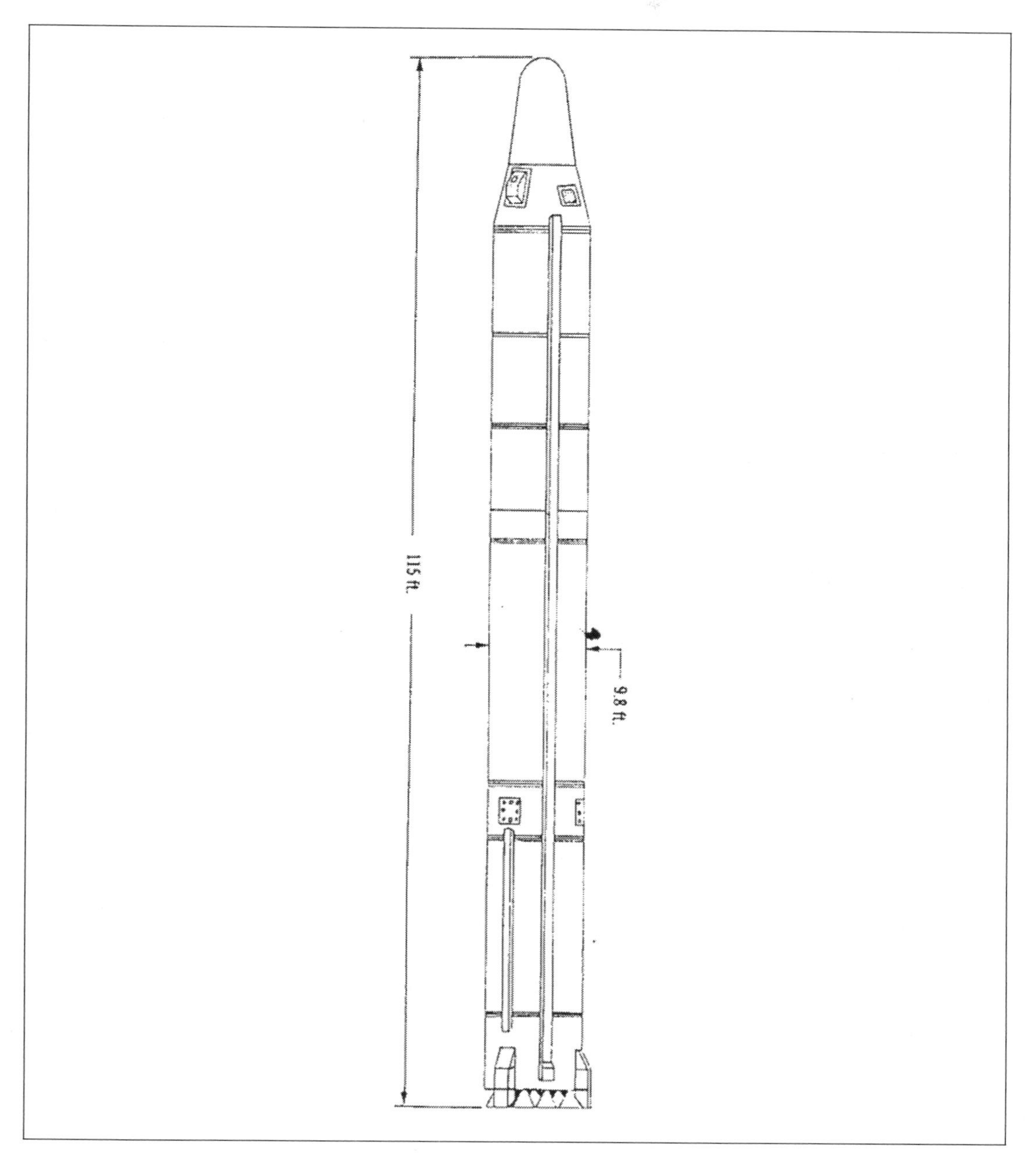

1960's NIE graphic showing the western intelligence inference on the appearance of the 8K67 ICBM from which the 8K69 FOBS was developed. The intelligence estimates included a length of 35.05 m (115 ft.) and a diameter of 2.98 m (9.8 ft.) whereas the true values were 32.2 m for length and 3 m for diameter. US Gov.

The 8K67 of the R-36 missile complex had its May Day parade debut on Red Square, Moscow, in May 1967.

As the various missile design bureau's in the Soviet Union had got down to the business of designing what were termed 'heavy' and 'super heavy' ICBM's, Yangel's OKB-586 (now Yuzhnoye State Design Office) put forward proposals for ballistic (ICBM) and orbital (FOBS & possibly MOBS) variants of the long range missiles of the R-36 missile complex that was being developed as a second generation Soviet ICBM. Such missile complexes were designed to be accommodated in single launch pad silos with a reaction time from a high alert status of 4 minutes, fully fueled and ready to launch missiles able to remain on silo alert for a period of up to seven years with a single maintenance period every three years.

The initial intention appeared to be to strike a target after less than one complete orbit, i.e., a fraction of an orbit, in which it would fit nicely with the FOBS label. However, the warhead could theoretically conduct a large number of orbits before the deorbit to attack a specific target area, this placing the weapon, the flight range of which was described by Yuzhnoye State Design Office as "unlimited", into an orbital weapon category, but still distinct from the type of weapon system that would be characterized as a MOBS in that it was designed to be held on ground alert and then launched, as would be an alert ICBM. By contrast, the MOBS as has been noted, was characterized as a weapon system that was placed into an Earth parking orbit for an indefinite period, theoretically from several days to several years, in peacetime or periods of reduced tension. Such systems would be brought to readiness in periods of increased tension or in the event of hostilities between the power blocks of East and West.

The advantage of a FOBS weapon complex did not stop at the theoretically unlimited flight range, but included the ability to attack targets from what is often described as any direction, which was of course theoretically true. However, the

FOBS could more realistically be described as a weapon system with excellent attack capabilities from two separate directions, stated by Yuzhnoye State Design Office as "front and rear". The directions of attack could, however, be better defined, as far as being deployed against targets in the CONUS (Continental United States) was concerned, as North and South, as opposed to the inflexible one direction attack profile inherent in ICMB's. This placed potential adversaries, in this case, as noted above, the target being undoubtedly the CONUS, at the disadvantage of having to provide warning and missiles defenses to counter warheads attacking from more than one direction. This contrasted with normal ICBM trajectories that would adopt the over the Arctic and North Polar regions profile facilitating the less complicated provision of capabilities to warn against attack by such missiles.

R-36 missiles on transporter dollies in Moscow circa 1967.

The depressed trajectory mode, generally termed DICBM (Depressed-trajectory Intercontinental Ballistic Missile), that was also inherent in the FOBS, and indeed potentially in a MOBS, would further complicate warning efforts, which would, in regards to detection and reaction times for an incoming attack, be reduced. This was certainly the case with the 8K69 weapon complex which was designed to fly specific reentry flight profiles, whether flown on an orbital or fractional-orbital strike profile, of low-angle trajectories specifically designed to further reduce or eliminate warning of the incoming warhead and, to a small degree, reinstate some of the accuracy that would normally have been lost in an operational FOBS.

The missile complex could also fly a DICBM flight profile distinct from the FOBS mode in that it would be launched on a ballistic trajectory that would not achieve Earth orbit in a similar mode to that of an ICBM, but with a considerably lower apogee to that normally achieved by an ICBM.

R-36 missile stages under construction and test. The first and second stages were universal for all variants of the R-36 missile complex and the Cyclone-2 and Cyclone-3 space payload launch vehicles. Yuzhnoye State Design Office

The 8K69, described by Yuzhnoye State Design Office as an orbital weapon and by western nations as a FOBS, both terminologies being accurate to a certain degree, was developed under the Soviet Union's orbital weapons programs more or less concurrently with the 8K67 ICBM from which it was derived. While its rivals at EDB/OKB-52 and Affiliation 1 of the Machine Building Central Design Bureau (now Salyut Design Bureau, a part of Krunichev Space Centre Company) had been developing the UR-200, and OKB-1 was developing the GR-1, OKB-586 was heavily committed to the development of the 8K67 that would be developed as the two-stage, 'tandem', single warhead ICBM's with the NATO designations SS-9 Mod 1 & SS-9 Mod 2. As with all second generation Soviet ICBM's under development in the 1960's, the 8K67 was designed with leak tightness levels for the liquid propellant storage compartments that would meet the required 7-year storage period for a fully fueled missile, a number of first generation missiles having this capability introduced as a modification.

The engines of the R-36 family of missiles burned UDMH (Unsymmetrical Dimethylhydrazine) fuel and NTO (Nitrogen Tetroxide) oxidizer, the 8K67 being the first missile to use this type of oxidizer which proved to be more dynamic than those previously utilized, such as the AK-27I oxidizer used in the 8K65 (R-14) ICBM. The introduction of UDMH fuel had commenced with the R-14 missile complex, replacing the TM-185 fuel used in 8K63 missile of the R-12 complex.

Testing of the R-36 stage engines at what is now Yuzhnoye State Design Office, Ukraine. Yuzhnoye State Design Office

Nose section atop an 8K67 ICBM in the Museum of the Strategic Missile Forces. MODRF

The first launch of an 8K67 ICBM took place on 28 September 1963, a further 84 developmental and service test launches being conducted during the period 1963-1966, for a total of 85 launches, 83% of which were deemed to have been successful. Including the development test launches mentioned above, and the service test launches, a total of 146 8K67 missiles were launched before the system was retired in 1978.

While the 8K67, in both the Mod 1 and Mod 2 configurations, was a single warhead missile, the 8K67P was developed as a MRV (Multiple Reentry Vehicle) variant of the missile able to accommodate three separate warheads. The first launch of this variant was conducted in August 1968, more than two and a half years after the first test launch of the 8K69 and almost 5 years after the first launch of an 8K67 ICBM.

The 8K67P introduced a number of improvements and capability enhancements to the ICBM element of the R-36 missile complex. Foremost among the changes was the aforementioned introduction of the MRV payload to allow more than a single warhead to be carried and the introduction of penetration aids to better allow the warheads to successfully strike the intended target. Modifications to the onboard control system were required to facilitate electronics coupling between the missile control system and the warhead payload in the nose section, as well as improvements to the ground based launch support equipment compared to that employed by the 8K67. There was, however no intentions or attempt to develop a FOBS derivative of the 8K67P which was deployed purely in a traditional ICBM mode.

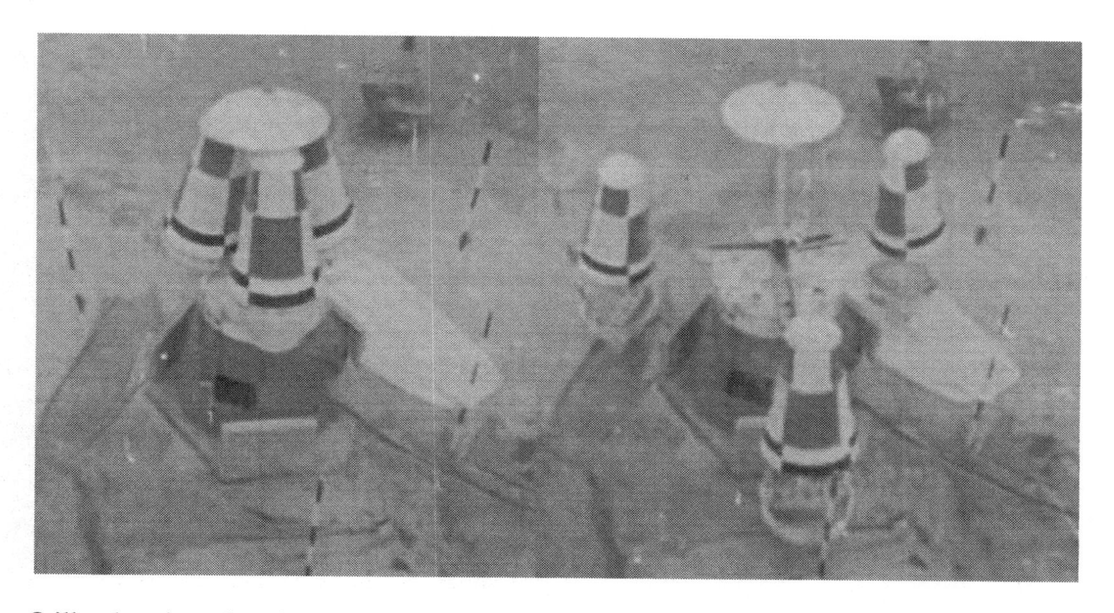

Stills showing the clustered multiple reentry vehicle warheads for the 8K67P in the stowed position (left) and in the spread for deployment position (right). Yuzhnoye State Design Office

8K69 FOBS/DICBM – The 8K69 (SS-9 Mod 3) is without a doubt an anomaly in regards to the deployment of ballistic missiles for strategic strike. As noted above, the system was referred to by the Soviet Union as an orbital weapon system and by the United States as a FOBS. The system was deployed as a partial counter to the NATO, in particular the United States, overwhelming advantage in being able to position significant numbers of MRBM/IRBM (Medium Range Ballistic Missile/Intermediate Range Ballistic Missile) on the periphery of NATO territory bordering Soviet or Warsaw Pact territory, whereas the Soviets, particularly following the Cuban Missile Crisis of October/November 1962, were restricted in their MRBM/IRBM deployments, being able to target against European and extended Eurasia target sets which would encompass China. The deployment of significant numbers of American MRBM/IRBM in Europe/Eurasia would reduce the warning time available for an incoming ballistic warhead, as well as increasing the number of attack directions available to NATO as opposed to the northerly (over the Pole, so to speak, direction of an ICBM attack).

The 8K69 adopted the first and second stages of the 8K67 that contained the respective propulsion systems for the RD-855 and RD-856 rocket engines that had been tried and tested in the 8K67 development phase. However, the 8K69, which unlike the other R-36 family variants, would be required to insert its payload into a low Earth orbit, added a third stage containing an RD-854 deceleration boost rocket engine for the orbital control of the orbital payload containing the nuclear warhead. This, along with the introduction of a control system centered on a system to vary the sustainer cut-off time and the ignition time of the deceleration propulsion system, being the major additions to the missile complex.

Development of the RD-855 and RD-856 steering engines for the R-36 missile complex first and second stages respectively had commenced at OKB-586 in 1961 when full-scale development of the 8K67 had commenced. The RD-855, used in the missile first stage, was developed as an open-cycle, four chamber single-burn rocket engine featuring a turbo-pump feed system, the turbine of which employed producer gas as its working fluid for the hypergolic propellants, ignition activating the turbo-pump assembly rotor, which was spun by the pyrostarters. The engine provided the thrust and controlled the flight of the missile in all stabilization axis. Stage flight control was conducted by gimballing each of the four chambers in one plane.

RD- 855 – data furnished by Yuzhnoye State Design Office

Engine mass: 320 kg
Propellants
Oxidizer: NTO (Nitrogen Tetroxide)
Fuel: UDMH (Unsymmetrical Dimethylhydrazine)
Sea-level thrust: 29100 kgf
Specific impulse
Sea level: 254 kgf/s/kg
Vacuum impulse: 292 kgf/s/kg
Propellant mixture ratio: 1.97
Gimbal angle: ± 41 angular °
Running time in flight: Up to 127 seconds

The RD-856, used in the missile second stage, was, like the RD-855 used in the first stage, developed as an open-cycle, four chamber single-burn rocket engine featuring a turbo-pump feed system, the turbine of which employed producer gas as its working fluid for the hypergolic propellants, ignition activating the turbo-pump assembly rotor, which was spun by the pyrostarters. The engine provided the thrust and controlled the flight of the missile in all stabilization axes. As was the case with the first stage RD-855, stage flight control was conducted by gimballing each of the four chambers in one plane.

RD-856 – data furnished by Yuzhnoye State Design Office

Engine mass: 112.5 kg
Propellants
Oxidizer: NTO (Nitrogen Tetroxide)
Fuel: UDMH (Unsymmetrical Dimethylhydrazine)
Vacuum thrust: 5530 kgf
Vacuum specific impulse: 280.5 kgf/s/kg
Propellant mixture ratio: 1.98
Gimbal angle: ± 30 angular °
Running time in flight: Up to 163 seconds

R-36 missile complex first stage incorporating the RD-855 propulsion system. US DoD

The RD-855 and RD-856 were used not only in the first and second stages of the 8K67, 8K67P ICBM and the 8K69 FOBS, but were also retained as the engines of choice for the OKB-586 Cyclone-2 and Cyclone-3 space payload launch vehicles that were developed from the 8K69. Production commenced in 1962 for the 8K67 missile complex and ended in 1992 with deliveries of the last units for the Cyclone launch vehicles. In the second decade of the 21st century the RD-855 and RD-856 engines are being restored to production status for employment as the first and second stages of the Ukrainian Cyclone-4 space payload launch vehicle.

The 8K69 orbital weapon complex used an RD-854 deorbit propulsion system on the boost stage. This engine was used for the deceleration and control of the orbital payload, containing the nuclear warhead, in all of the stabilization axis.

Development of the open-cycle RD-854, which was designed to burn NTO and UDMH propellants, commenced at OKB-586 in 1962, this being the first Soviet rocket engine program to introduce, and place into production, a pipe nozzle for the engine chamber.

RD-854 – data furnished by Yuzhnoye State Design Office

Vacuum thrust	Propellants	Vacuum specific impulse	Mass	Missile
7700 kgf	NTO Oxidizer UDMH fuel	312.2 kgf/s/kg	100 kg	8K69 (SS-9-3)

Top: R-36 second stage incorporating the RD-856 propulsion system complete with lateral mounted gimballing nozzles. Above: The launch of an 8K67P ICBM from a missile silo. Such launch characteristics were common to all R-36 derivatives including the 8K69. Yuzhnoye State Design Office

Reentry Vehicle/Warheads – The RV (Reentry Vehicle) segment of a ballistic missile is basically a capsule of sorts designed to house the warhead, which is protected from the exoatmospheric and atmospheric forces of nature, including, in regards to the latter, the intense heat generated during the reentry into the Earth's atmosphere from the low-Earth orbital environment. The R-36 missile family employed a traditional blunt front end RV design for their respective nuclear warheads, which, in the case of the 8K69, was basically part of the overall orbital vehicle that would have been equipped with an INS (Inertial Navigation System) and radio altimeter. In imagery the OGCh 8F021 warhead appeared less blunt due to its smaller size in comparison to the larger warheads of the 8K67. Warhead weight was in the order of 1678 kg (3,700 lb.), this being the major element of the missile throw weight that was normally defined as the portion of the system located above the last booster stage. The throw weight, which it is estimated was similar to the 3-5 metric tons, achieved by the Cyclone-2 space launch vehicle, would include the orbital/RV, encompassing the warhead, which contained the system to facilitate the retrofire, or deorbit burn, or to deboost such a vehicle from its normal ballistic trajectory. No definitive value has been provided for warhead yield, but this was considered to be in the region of 2+ to 5 MT, considerably lower than the 8F675 warheads arming the 8K67 Mod 1 and Mod 2 which were in the order of 8 and 20 MT yield respectively.

In the 1960's, the Soviets began developing sharper front ends for their warhead reentry vehicles, moving away from the blunter front ends employed in first and second generation systems. This was designed to increase accuracy by increasing the ballistic coefficient characteristics, determined as the weight of the RV divided by the RV area and drag coefficient. In other words the sharper front of the RV would have a tendency to move at higher velocities though the upper atmosphere. The faster transit contributing to reduced forces of deflection, therefore, increasing accuracy.

Sharper front end warhead RV were expected, in the early 1970's, to reduce CEP (Circular Error Probability) to around 0.46 km (0.25 nm) for a ballistic trajectory ICBM, although this would be expected to increase for a FOBS employing such RV technology, there, however, being no program for re-equipment of either the 8K67 ICBM's or the 8K69 FOBS with the new RV designs.

There was no right or wrong on the RV ballistic coefficient, each facilitating a specific operational priority. A RV possessing higher ballistic coefficient would itself be less susceptible to potential accuracy errors that could result from being subjected to the effects of natural phenomena such as wind and atmospheric density. A RV possessing lower ballistic coefficient, while being more susceptible to the above natural phenomena, would be less susceptible to non-natural phenomena such as the blast effects, wind, dust and debris, of recent nuclear bursts in the target area, of some importance for missiles tasked with attacking enemy missile silos. In addition, lower ballistic coefficient RV's were more easily hardened against defensive interference such as the radiation effects of a nuclear armed ABM system of the type that was being developed in the United States in the 1960's.

The later 8K67P ICBM MRV consisted of three 8F676 warheads, with yields in the order of 2+ MT each, housed in a smaller RV than the single warhead carried by the 8K67. The main logic behind the MRV concept, which would involve the release of the individual RV's when in either the late free flight or the terminal (reentry) phase of the attack, was to confuse enemy radar systems that would be controlling a potential ABM defense system, thus aiding the ability of the weapon to successfully strike the selected target. A secondary consideration, or, depending on operational requirements, the primary consideration, was to increase the area affected by the attack, particularly if employed against soft targets such as cities, ports or industrial centers. The MRV were distinct from the MIRV (Multiple Independently targeted Reentry Vehicle) in that the latter were able to be targeted against separate targets while the former were targeted against the same general target area.

8K67P MRV's are thought to have been flight tested from Tyuratum, Baikonur cosmodrome, on 23 August and 11 September 1968, the payload of three MRV's impacting the target range area on the Kamchatka Peninsula in the Soviet Far East, a firing range in the order of 6296 km (3,400 nm). There was, however, no plan to introduce a MRV capability to the 8K69 which retained the single warhead configuration throughout its service life.

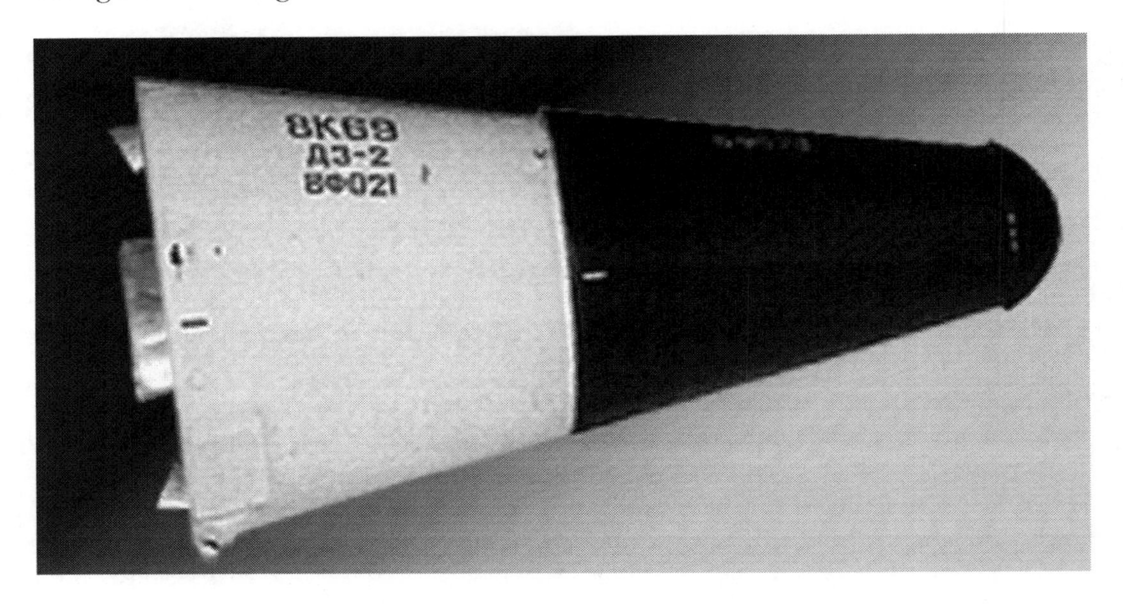

The 8K69 complex was armed with the 8F021 nuclear warhead which had a mass of some 1678 kg and a projected yield somewhere between 2+ and 5 MT.

NIE 11-8-68, dated October 1968, lent weight to the assessment that NATO was unclear exactly what type of system the Soviets were developing, FOBS, DICBM or perhaps a duel system capable of operating in both modes. What was clear was that the system had the potential to strike undetected, or at least with considerably reduced warning time, by circumventing the existing BMEWS (Ballistic Missile Early Warning System). Intelligence document NIE 11-8-72 concluded that the 8K69 complex (referred to in the document as the SS-9-3) could attack targets "negating or delaying detection" by the US BMEWS. Although it is often stated that the successor US warning systems removed this capability, the intelligence assessments conclude only that such warning systems promised to reduce the advantage. Later improved early warning systems would, to a certain degree, reduce the amount of time a detected incoming warhead had before impact, but this was largely a moot point as the ABM system being developed in the 1960's had long since been abandoned due to the ABM Treaty that limited the deployment of such systems, which would therefore prove ultimately unacceptably expensive to deploy in small numbers.

Command and Control – The operational control system for a FOBS would be along the same lines as that required for an ICBM/DICBM in that it would be targeted prior to launch. Additional requirements placed on the command and control system would be necessary for the placing of the payload into a low-Earth orbital trajectory and for the deorbit maneuver burn. If used as a FOBS system, the warhead would be targeted to strike targets before one complete orbit of the Earth had been completed. However, if required the targeting process could have allowed for multiple orbits to be completed, there being obvious advantages in the targeting coverage available in such a case, counterbalanced by the obvious disadvantage of increased likelihood of detection of the RV prior to the terminal reentry attack phase.

The command and control for the 8K69 force would have been provided by military tracking systems already in place for the existing Soviet ICBM command and control network. The system would also have benefited from the extensive Soviet space object tracking network developed to track near Earth spacecraft and satellites out to lunar distances.

The Soviet command and control network for ICBM and spacecraft in Earth orbit had its early beginnings following the 1955 decision to build the ICBM test facility now part of what is known as Baikonur Cosmodrome in Soviet Kazakhstan (now and independent republic). This KIK (Ground Control Station) attained operational capability in 1957, the year of the first successful ICBM launch and the first artificial satellite to be placed in Earth orbit. Construction commenced for an ICBM test range in the Arkhangelsk district (the present day Plesetsk Cosmodrome) in 1957, and in 1960 the 3rd Department of the Main Missile Directorate of the Ministry of Defense of the USSR (Union of Soviet Socialist Republics) was formed to facilitate the organization of space control This was formed into the Central Spacecraft Directorate of the USSR Defense Ministry in 1964.

The 8K69 missile employed an ICS (Integrated Control System) similar to that of the 8K67 ICBM. Such systems had no further link to a ground control center following the missile launch, this bestowing a clear distinction between the FOBS and the MOBS concept, allowing the former to be exempt from the treaties proscribing the deployment of orbital weapons.

The ICS was basically made up of two main components – the accelerometer and a gyroscope. The accelerometer was basically an instrument that would provide measurements of the acceleration of the missile in the direction it had been launched. It is unclear how many accelerometers were incorporated, but it widely accepted that three such instruments that were mounted at the appropriate right angles to each other would provide full coverage of the host missiles acceleration profile during the powered phase of the flight.

The function of the gyroscope was primarily to provide measurements of the deviation of the missile from its reference direction during flight. It is unclear how many gyroscopes were incorporated, but, as was the case with the accelerometers, it is widely accepted that three such instruments that were mounted at the appropriate right angles to each other would provide accurate measurements of any deviation from the reference direction during the missile powered flight phase.

Ranges attained by the 8K69 (SS-9 Mod 3) in the DICBM mode tests were estimated by western intelligence agencies as being up to 11667.6 km (6,300 nm) according to intelligence documents, which, as stated quite emphatically in NIE 11-8-72, could "unquestionably provide full coverage of the US [CONUS] on northerly trajectories from any SS-9 site", that naturally included Tyuratum where the 8K69 was deployed. However, the NIE 11-8.72 intelligence assessment went on to imply that the SS-9 Mod 3 (8K69), when operated in a FOBS mode, would not be "capable of inserting the payload into an orbit that would permit an attack against any target in the US on the initial orbit, on either northerly or southerly orbits". There was of course a large degree of uncertainty in regards to this statement, which, in regards to a southerly trajectory launch, would most likely have been inaccurate, their being a high degree of confidence that targets in the CONUS could be reached on the initial orbit, particularly with regards to coverage of targets located in the eastern third of the country when the Mod 3 was launched on a southerly trajectory from the Dombarovskiy missile complex. Of course, as previously noted the Mod 3 was deployed only at Tyuratum. The assessment went on to conclude that targets in the eastern one third of the CONUS could be attacked by the FOBS if launched in a southerly direction from the SS-9 complex at Dombarovskiy. However, it added the clear caveat that it was based on limited knowledge of the system and questioned the data that suggested the Soviets were deploying a system of only limited capability.

One idea was that the system was to be deployed operationally only in the DICBM mode and that the tests in the FOBS mode were conducted purely on a research basis with no intention to develop or deploy the system as such. Another idea was that it would be deployed primarily as a DICBM and that the Soviets were taking advantage of the inherent FOBS capability, which would be limited. In 1967, a third possible course of action that could have bene taken by the Soviet's was put forward, in that the system could be modified later to provide additional range that would permit the system to attack targets throughout the entire CONUS in the FOBS role. Of course, the system as developed and deployed was apparently capable of attacking all CONUS targets in the FOBS role.

In understanding NATO's underestimation of the capabilities of the 8K69 FOBS/DICBM, it has to be considered that in the 1960's and 1970's the various western intelligence agencies had, to varying degrees, underestimated the capabilities of the entire R-36 missile family, which in most assessments was woefully short of the respective missiles actual capabilities. This underestimation was the catalyst for the misinterpretation of the SS-9 Mod 3 (8K69) capability in the both the DICBM and FOBS modes.

Intelligence estimates drew too much on the SS-9-1 (8K67 Mod 1) range of firings of 13149 km (7,100 nm), reduced to about 12223 km (6,600 nm) when the Coriolis Effect (sometimes referred to as the Coriolis force, although no actual force is present) of the Earth's rotation was taken into account. This would be referred to as a NRE (Non-Rotating Earth) range that was reached by subtracting the 500 nm estimated to have been added due to the Earth's rotation on its axis as it orbits the Sun.

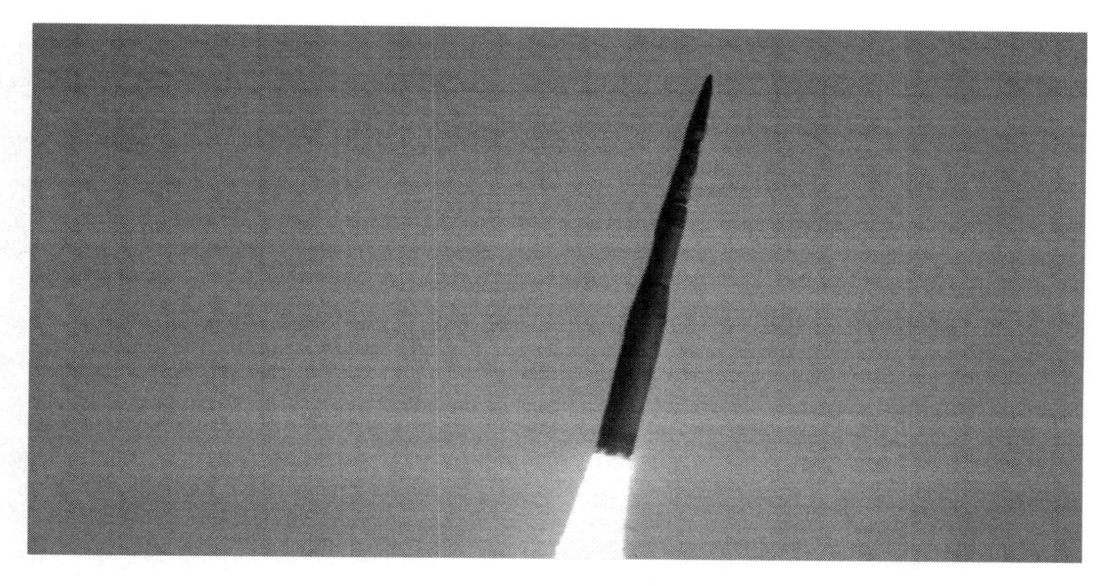

It is doubtful that the 8K69 would have been targeted against the US ICBM force which was housed in hardened missile silos (an LGM-30B Minuteman I shown) as it most likely lacked the accuracy for such a role. US DoD

From a northern hemisphere perspective the Earth rotates on its axis in an anti-clockwise direction – to the East. The Coriolis Effect creates what is termed an unbalanced force which itself lends to increased acceleration with the result that the prevailing winds are deflected to the right as the Earth rotates on its axis, the same effect being exerted on manmade objects such as a ballistic missile.

Taking the Coriolis Effect into consideration means that while the intended track of a missile launched from a high northern latitude is in a straight line, the apparent track bends to the right in line with the rotation of the planet as the Earth rotates around the north-south axis. Therefore, winds blowing in the northern hemisphere have an apparent movement to the right whilst winds bowing from the southern hemisphere have an apparent movement to the left. This apparent movement, the Coriolis Effect, increases in intensity with the increase in latitude. In effect, the intended path of an object, in this case a ballistic missile, is deflected by the Coriolis Effect due to the Earth's rotation. That said, the test launches of the 8K69 in both the DICBM and FOBS modes were conducted from a West to East direction more or less in concert with the rotation of the planet, vastly reducing the Coriolis Effect in regards to deflection from the intended track to the apparent track. To a small degree the effect of the Earth's rotation would have assisted the 8K69 in achieving orbital velocity. The intelligence gaff here was in the inference that the West to East launches were required to allow the missile to achieve orbital velocity, which was apparently not the case, said direction of launch being in order to deorbit the test warhead vehicle on the range in the Soviet Far East after only a fraction of an orbit when tested in the FOBS mode.

The 1968 intelligence estimates included a maximum NRE range of 12223 km (6,600 nm) for the SS-9-1 (8K67 Mod 1), this being the observed test flight distance as opposed to the actual value of 15200 km (8,207 nm) achievable by the missile. The same years estimate for the SS-9-2 (8K67 Mod 2) was 9260 km (5,000 nm) as opposed to the actual value of 10200 km (5,507 nm) achievable by the missile. By 1972 these estimates had increased to 12964 km (7,000 nm) for the SS-9-1, still in excess of 2000 km short of its true operational range. Even with NATO's under estimation of the actual range capability of the SS-9-1, this weapon was deemed to be capable of striking any target in the CONUS from any of the known SS-9 launch complexes.

NATO continued to vastly underestimated the range capability of the SS-9-2 (8K67 Mod 2), which by its observations it deemed had not been flown to greater ranges than 8148 km (4,400 nm) whereas the missile had an actual operational range of 10200 km. The intelligence assessment NIE 11-8-72 wrongly inferred that this variant could only target "the extreme northwestern part of the United States from the closest SS-9 complex". After using the unreliable process of extrapolation (unknown values being inferred), it was eventually concluded by US intelligence agencies, outlined in NIE 11-8-72, that the SS-9-2, employing a minimum energy trajectory, had a maximum range of 9630 km (5,300 nm), which was still almost 600 km short of the actual value of 10200 km. The underestimation left the intelligence estimates to mistakenly infer that these missiles, while being able to target all six of the CONUS Minuteman ICBM complexes, could only target a single Titan ICBM missile complex, as well as NORAD (North American Air Defense) and SAC (Strategic Air Command) HQ's from a single SS-9 complex, whilst the missile could in fact target more or less the entire range of strategic targets in the CONUS.

It should be noted that the DIA (Defense Intelligence Agency) was of the opinion that for the SS-9-2 a maximum range of 10186 km (5,500 nm), more or less the correct range value of the missile, should not be ruled out. This assessment went on to conclude that such a range would allow complete target coverage of all the US Minuteman complexes from any of the SS-9 complexes.

The range underestimations remained in force throughout the service lives of the SS-9-1 and SS-9-2, the true range of these missile complexes, which allowed complete coverage of CONUS targets, not being appreciated until the years following the end of the Cold War in the 1990's by which time they had long since been retired from operational service.

Before embarking upon a discussion of warhead accuracy it should be noted that the term CEP (Circular Error Probability) is not a reference to the projected or estimated accuracy of a single weapon, but rather a projected accuracy expected for

50% of an attacking missile force to fall within, it being concluded that the other 50% of the attacking force would deliver their warheads to within a CEP of 3.5% of the intended target area.

It was inferred, as outlined in NIE 11-8-72, within NATO that the SS-9-1 and SS-9-2 would be deployed against CONUS based ICBM silos. This was initially intended to cover the Titian II force, but later included the Minuteman ICBM silos. The SS-9-3 (8K69) FOBS was considered to lack the accuracy for effective use against hardened targets of the missile silo type, NATO's inference being that this system would be employed against soft targets such as cities, industrial centers, ports and some military infrastructure facilities and perhaps SAC bomber bases.

CEP estimates for the SS-9-1 and SS-9-2 were varied from agency to agency, no actual Soviet figures being released. These estimates ranged between lows of 0.74 km (0.4 nm) and 1.1 km (0.6 nm) for a missile launched to a distance of around 9815 km (5,300 nm). More detailed analysis, actually best estimates, laid down in NIE 11-8-68, were about 0.92 km (0.5 nm) when radio-inertial guidance was employed and about 1.39 km (0.75 nm) with inertial guidance only. These values were altered, following receipt of data obtained from Soviet testing, to between a low of 1.1 km (0.4 nm) and a high of 1.3 km (0.7 nm), both sufficient for targeting Minuteman silo complexes, the lower figure of course having a far higher chance of success.

In regards to the 8K69 (SS-9-3/Mod 3) FOBS, intelligence agencies widely regarded the CEP to be considerably higher than that of the SS-9-1 and SS-9-2 ICBM's. Values of around 2.78 km (1.5 nm) to 5.55 km (3.0 nm) were postulated for the missile when employed in a FOBS mode and launched on a southerly trajectory. Values of 1.85 km (1.0 nm) to 3.70 km (2.0 nm) were postulated for the missile when employed as DICBM (such a launch profile would be in a similar northerly direction to that adopted by standard ICBM targeting), no figures, of course, being released by the Soviet Union. With the increased CEP ranges the 8K69 would have been far less suited to attacking hardened targets such as Minuteman silos as the chances of knocking same out was vastly reduced compared to the SS-9-1 and SS-9-2 missiles, both of which also had considerably increased warhead yield over that of the 8K69. For this reason the 8K69 appears to have been far more suited to attacking soft targets as noted above.

The reduction in CEP was the price that had to be paid for the systems extra-ICBM qualities, the potential to strike targets with reduced detection time being of particular concern for NATO, the US in particular, as it allowed the FOBS to be used as a viable first strike weapon. In such a scenario the 8K69 would, perhaps, have been able to take out a significant portion of the SAC bomber fleet on the ground prior to a main attack which would see the SS-9-1 and SS-9-2, and other Soviet ICBM's, launched against the CONUS ICBM complexes.

Of course, if employed as a DICBM with a ballistic trajectory of considerably lower apogee to that of an ICBM, the accuracy shortfalls that were inherent in a 1960's technology FOBS would no longer be such an issue. Such trajectories could have been used to increase the accuracy of the warhead, although, perhaps counter-intuitively, western intelligence agencies inferred that such a system would be less accurate than a warhead delivered in a normal ICBM flight profile, intuition of course pointing to the opposite, flight range being the factoring value. The drawback for the DICBM would be a considerable reduction in maximum operational range over that attained in a conventional (so to speak) ICBM flight trajectory profile.

The first launch of an 8K69 missile was conducted in December 1965, there being a total of 19 developmental and test launches of the complex, which entered service with the Soviet Union Strategic Missile Force in 1969 and was retired in 1983.

Graphic depicting the three missile variants that constituted the R-36 second generation strategic strike system. Yuzhnoye State Design Office

R-36 Missile Family – 8K67 ICBM, 8K69 FOBS, 8K67P ICBM – data furnished by Yuzhnoye State Design Office

	8K67	8K69	8K67P
Launch weight:	183.9/182.0 tnf	181.297 tnf	183.45 tnf
Length:	32.2 m	32.65 m	32.2 m
Diameter:	3 m	3 m	3 m
Propellants	Liquid, high-boiling hypergolic		
Oxidizer:	NTO	NTO	NTO
Fuel:	UDMH	UDMH	UDMH
Number of warheads:	1	1	3
Maximum operational Range:	10200/15200 km	Unlimited within one Earth orbit	10200 km
In service:	1967-1978	1968-1983	1970-1979

3

8K69 ORBITAL MISSILE COMPLEX DEVELOPMENT & SERVICE TEST LAUNCHES

The first ballistic missiles to enter operational service in the Soviet Union were effectively operated by artillery units of the Soviet Army. However, the increasing ranges that would become available for such weapons would take operational concepts for ballistic missiles beyond the battlefield, or indeed, beyond the immediate regional conflict zone, justifying their removal from army command (battlefield ballistic missiles such as the R-11 series would remain under Soviet ground forces ownership and command). To this end, the SMF (Strategic Missile Force) was formed as an independent entity of the Soviet Armed Forces by the decree of the Council of Ministers of the USSR (Union of Soviet Socialist Republics), No. 1384-615, dated 17 December 1959. This organization absorbed the strategic and some non-strategic ballistic missile formations then in existence.

All ICMB (Intercontinental Ballistic Misisle) and the 8K69 orbital weapon would be operated by the SMF, the introduction of the 8K67 in 1967 introducing a quantum leap in capability for the force as it was considered a viable weapon in terms of accuracy and certainly in warhead yield for the targeting of so called hard targets such as the protected silo complexes for the US Minuteman light ICBM force. It is noted that while US intelligence estimates repeatedly put the year of IOC (Initial Operational Capability) for the SS-9-2 (Mod 2) as 1966, the Mod 1 achieving IOC the following year, Yuzhnoye State Design Office documentation states emphatically that the IOC for both variants was attained in 1967.

In total there were nineteen development and service test launches of the 8K69 missile complex in either DICBM (Depressed Trajectory Intercontinental Ballistic Missile) or FOBS (Fractional Orbit Bombardment System) mode, commencing with the first in December 1965. A further two launches were conducted under unknown designations prior to the launch designated Cosmos 139 on 25 January 1967. Of the 19 test launches conducted, 89% were deemed to have been successful by Yuzhnoye State Design Office, the OKB-586 successor organization.

The nineteen launches, fifteen of which have a known Cosmos designation, are listed below in chronological order of launch date. Note: There have been several, fallacious, claims that list as many as 24 test launches, however, Yuzhnoye State Design Office documentation clearly states the figure of 19 launches and US agencies, were able to monitor only 18 of these launches, 15 of which were monitored by NASA (National Aeronautics and Space Administration). The remaining launches were not part of the 8K69 development program and, without validated evidence to the contrary, have therefore been excluded here. Some of these lists, most, if not all, of which appear to take their information from a single source, do not list the last two known 8K69 launches, which, when added to the inaccurate value of 24 gives an new inaccurate value of 26.

Intelligence document NIE 11-8-68, backed by former Soviet sources, stated that the "initial tests", covering the first few launches, of the 8K69 system were conducted in a flight profile having an apogee of around 120 nm. This would certainly be suggestive that the initial development testing focused on validating the DICBM mode, the missiles being launched from Tyuratum with the warheads impacting in the Kamchatka target area (normal apogee for an ICBM flown to the ranges involved would be in the order of 400-500 nm). That said, in at least two of the early launches (pre-September 1966) a true ballistic trajectory was not flown by the RV (Reentry Vehicle), which, in both cases, certainly appears to have undergone an alteration of trajectory through de-orbit retrofires. NIE 11-8.68 credits a test flight in September 1966 as the first of the true FOBS mode flight profile tests.

There was a single flight test in January 1967, followed by an unsuccessful test in March and another in May, paved the way for a concerted flight test effort commencing on 17 July and continuing through October 1967. NIE 11-8-68 stated that seven FOBS tests took place between July and the end of October 1967, this being borne out by NASA records of FOBS tests conducted under the Cosmos designations during the same period. All of these launches, which were successful, were conducted in an easterly direction, which, although this would certainly have provided the not insignificant benefit of the Earth's rotation to aid velocity for placing the payload into a low Earth orbit, the inference by western intelligence agencies that this was the main reason for the easterly trajectory was probably wrong. The true reason in all likelihood was simply that from the Tyuratum launch complex, the target area for missile launches was easterly toward the Kamchatka peninsula. A launch in a northerly direction or southerly direction would probably not have been able to be de-orbited over Soviet territory on a fractional orbit profile.

These tests were among the main reasons that US intelligence estimates inferred that the system could only strike targets in the US after one or more complete orbits of the Earth, which would reduce the systems overall effectiveness to strike its targets without being detected or with much reduced detection time, but not its effectiveness to circumvent the existing ABM defenses (it should be noted that the rudimentary 1960's ABM defenses were highly unlikely to achieve much success in the event of a full-scale nuclear exchange).

Following the 28 October 1967 flight test, there followed a lull of almost eight months before the next flight test of the system in the FOBS mode, this being conducted on 25 April 1968. The next two flight tests were then apparently flown in an DICBM mode, although the range of the flights, which followed ballistic trajectories with an apogee in the region of 300 nm (for a normal ICBM flight of the range flown an apogee of around 700 nm would have been considered more the norm), was increased over that of earlier flights. The RV's were brought down into the Pacific Ocean some 13705 km (7,400 nm) from the launch site at Tyuratum following a reorientation and ignition of the de-boost stage in the later stages of the flights. The missile second stages continued to fly on ballistic trajectories until impacting in the Pacific Ocean around 741 km (400 nm) downrange from the RV.

Following these tests the intelligence community within NATO began to infer the flight data as that of a DICBM featuring a "retrofired RV", NIE 11-8.68. Such a system, it was clear, would have restored a measure of the accuracy that had been forfeited by having to fly in a lower flight trajectory than a normal ICBM flight profile. No sooner had the western intelligence community came to a general consensus that any deployed system would indeed be a DICBM when a further flight test, conducted on 2 October 1968, was flown in the FOBS mode, throwing no little amount of confusion back into the mix in regards to what the system under development would emerge as. The various intelligence communities from this point had a much lower confidence in their ability to understand what the new weapon system would be. Perhaps, for the first time the realization emerging that the Soviet Union was, perhaps, developing a missile complex capable of being deployed operationally as a DICBM and as a FOBS.

8K69 Orbital Weapon Complex Development & Service Test Launches

First launch – 16 December 1965		
Undesignated launch – 19 May 1966	NIE 11-1-67 stated as last of 3 sub-orbital flights indicating 1 prior unacknowledged flight	
Undesignated – September 1966		
Cosmos 139 – 25 January 1967	NASA/NIE 11-1-69	
Undesignated – 22 March 1967	NIE 11-1-67/NIE 11-169	Failure
Cosmos 160 – 17 May 1967	NASA/NIE 11-1-69	Failure
Cosmos 169 – 17 July 1967	NASA/NIE 11-1-69	Success
Cosmos 170 – 31 July 1967	NASA/NIE 11-1-69	Success
Cosmos 171 – 8 August 1967	NASA/NIE 11-1-69	Success
Cosmos 178 – 19 September 1967	NASA/NIE 11-1-69	Success
Cosmos 179 – 22 September 1967	NASA/NIE 11-1-69	Success
Cosmos 183 – 18 October 1967	NASA/NIE 11-1-69	Success
Cosmos 187 – 28 October 1967	NASA.NIE 11-1-69	Success
Cosmos 218 – 25 April 1968	NASA/NIE 11-1-69	
Cosmos 244 – 2 October 1968	NASA/NIE 11-1-69	
Cosmos 298 – 15 September 1969	NASA	
Cosmos 354 – 28 July 1970	NASA	
Cosmos 365 – 25 September 1970	NASA	
Cosmos 433 – 9 August 1971	NASA	

8K69 flight tests conducted under the Cosmos designation and monitored by NASA

Cosmos 139: Cosmos 139 was launched from Baikonur Cosmodrome with an epoch start of 13.55.00 UTC on 25 January 1967 on what NASA inferred as a Tsklon launch vehicle. The orbital parameters included a periapsis of 144 km, apoapsis 210 km, period 87.5 minutes, inclination 50° and eccentricity 0.00503.

Cosmos 160: Launched from Baikonur with an epoch start of 16.04.00 UTC on 17 May 1967, orbital parameters being periapsis 137 km, apoapsis 177 km, period 88.4 minutes, inclination 49.6° and eccentricity 0.00306.

Cosmos 169: Launched from Baikonur with an epoch start of 16.48.00 UTC on 17 July 1967, orbital parameters being periapsis 135 km, apoapsis 200 km, period 87.6 minutes, inclination 50° and eccentricity 0.00496.

Cosmos 170: Launched from Baikonur with an epoch start of 16.48.00 UTC on 31 July 1967, orbital parameters being periapsis 121 km, apoapsis 252 km, period 88.1 minutes, inclination 49.4° and eccentricity 0.00997.

Cosmos 171: Launched from Baikonur with an epoch start of 16.04.00 UTC on 8 August 1967, orbital parameters being periapsis 138 km, apoapsis 177 km, period 87.5 minutes, inclination 49.6° and eccentricity 0.00298.

Cosmos 178: Launched from Baikonur with an epoch start of 14.52.00 UTC on 19 September 1967, orbital parameters being periapsis 138 km, apoapsis 258 km, period 88.3 minutes, inclination 49.6° and eccentricity 0.00912.

Cosmos 179: Launched from Baikonur with an epoch start of 14.09.00 UTC on 22 September 1967, orbital parameters being periapsis 139 km, apoapsis 207 km, period 87.8 minutes, inclination 49.6° and eccentricity 0.00519.

Cosmos 183: Launched from Baikonur with an epoch start of 13.26.00 UTC on 18 October 1967, orbital parameters being periapsis 130 km, apoapsis 315 km, period 87.4 minutes, inclination 50° and eccentricity 0.014.

Cosmos 187: Launched from Baikonur with an epoch start of 12.12.00 UTC on 28 October 1967, orbital parameters being periapsis 143 km, apoapsis 301 km, period 87.4 minutes, inclination 50° and eccentricity 0.01196.

Cosmos 218: Launched from Baikonur with an epoch start of 00.43.00 UTC on 25 April 1968, orbital parameters being periapsis 123 km, apoapsis 162 km, period 87.2 minutes, inclination 49.5° and eccentricity 0.00299.

Cosmos 244: Launched from Baikonur with an epoch start of 13.40.00 UTC on 2 October 1968, orbital parameters being periapsis 134 km, apoapsis 158 km, period (undetermined), inclination 50° and eccentricity 0.01184.

Cosmos 298: Launched from Baikonur with an epoch start of 16.04.00 UTC on 15 September 1968, orbital parameters being periapsis 127 km, apoapsis 162 km, period 87.3 minutes, inclination 49.6° and eccentricity 0.00268.

Cosmos 354: Launched from Baikonur with an epoch start of 22.04.00 UTC on 28 July 1970, orbital parameters being periapsis 144 km, apoapsis 208 km, period 87.5 minutes, inclination 50° and eccentricity 0.00488.

Cosmos 365: Launched from Baikonur with an epoch start of 12.12.00 UTC on 25 September 1970, orbital parameters being periapsis 144 km, apoapsis 210 km, period 87.7 minutes, inclination 49.5° and eccentricity 0.00503.

Cosmos 433: Launched from Baikonur with an epoch start of 22.44.00 UTC on 9 August 1971, orbital parameters being periapsis 159 km, apoapsis 259 km, period 89.3 minutes, inclination 49.5° and eccentricity 0.00758.

An 8K69 missile is prepared for launch during the developmental flight phase.

Estimates for SS-9 (covering all variants of the R-36 missile complex) numbers deployed in 1975 was put at 282, dropping to 264 the following year. Of these totals, eighteen were 8K69 (SS-9 Mod 3). Total estimated Soviet ICBM deployment around this time was in excess of 1500, the deployed force of eighteen 8K69 missiles representing less than 2% of the total combined force of ICBM/FOBS. It should be noted that a proportion of the above mentioned missiles would have been expected to have been, at least temporarily, out of service due to modifications and upgrades of systems and facilities. In this regard, at least 18 SS-9 silos of all Mods were considered to be offline in early 1975.

Development test launch of an 8K69 from Baikonur cosmodrome.

NATO assessments inaccurately concluded that the 8K79 was deployed at a single group of six silos at Tyuratam, Baikonur in Soviet Kazakhstan. However, while Tyuratum was indeed the base of choice, the operational force was housed in eighteen silos, IOC being attained in 1969, apparently the 25th of August, although the first missiles were apparently prepared for deployment the previous November (Yuzhnoye State Design Office puts ICO in 1968)) when the weapon system was accepted for operational service. It is inferred from available information that the full complement of missiles had been received at Tyuratum by 1971, serving until withdrawn from service in January 1983 to make the numbers available for new ICBM that were limited under START II (Strategic Arms Limitation II).

As was the case with any weapon system developed as a nuclear deterrent, with the explicit purpose of dissuading a would be aggressor from attacking the holder of that deterrent with nuclear weapons, then the success of the 8K69 orbital weapon can be gauged from the fact that it, like all nuclear weapons deployed during the Cold War, was never used in its designed role in anger, this, of course, being the true measure of success for all weapon systems developed as a deterrent. The 8K69 orbital weapon system had served operationally for just over fourteen years, being phased out of service in that same year that the Cold War reached its most dangerous hour with the little known war scare of 1983, a time when we, as human beings, truly were mad in line with the aptly used acronym MAD (Mutually Assured Destruction), but reason and common sense somehow prevailed.

4

8K69 ORBITAL MISSILE LEGACY – THE CYCLONE-2 & CYCLONE-3 SPACE PAYLOAD LAUNCH VEHICLES

Development of the orbital weapon complex placed OKB-586 in a good position to develop a lightweight space launch vehicle based on the 8K69 missile. Development of the resultant 11K69 Tsyclone-2 (Cyclone-2) two-stage launch vehicle (allocated the western space launch vehicle designation SL-11) commenced in 1965. This vehicle was designed to place payloads into circular, and elliptical low Earth orbits that would include what was referred to as open orbits.

The Cyclone-2 was apparently the first rocket system, certainly within the Soviet Union, designed with what is described by Yuzhnoye State Design Office as a "fully automated launch operations cycle", that required an extensive development period. The first Cyclone-2 launch was conducted from Baikonur Cosmodrome on 6 August 1969, there being a total of 106 launches of the system during its operational life with the Soviet Union and later the independent Ukraine, the last of which was conducted in June 2006. The Cyclone-2 had unprecedented and unrivalled reliability with all 106 launches being successful (the value of 106 launches was provided by Yuzhnoye State Design Office while the National Space Agency of Ukraine puts the number of launches at 116, all of which were stated as being successful).

As well as more traditional orbital payloads the Cyclone-2 also took on a military function in that it was adopted as the launch vehicle for the TMBDB Counter-space Defense Satellite Fighter, the only dedicated anti-satellite system to be operationally deployed. This system, which was allocated to alert duties in 1973, attaining full operational capability in 1978, served beyond the break-up of the Soviet Union in December 1991, being decommissioned by the Russian Federation in 1993.

The limitations of the payload lifting capability of the Cyclone-2 to medium and elliptical Earth orbits led to development of the three-stage 11K68 Cyclone-3 launch vehicle, development of which commenced in 1970 on the basis of the Cyclone-2 launch vehicle/8K69 missile. The Cyclone-3 third stage was, as detailed by Yuzhnoye State Design Office, "designed in a pressurized ('ampulized') version based on the Yuzhnoye engine capable of duel burn in zero gravity".

Top: Cyclone-2 prior to launch. Above: The Cyclone-2 and Cyclone-3 launch vehicles employed the RD-855 and RD-856 first and second stages of the R-36 missile complex. Yuzhnoye State Design Office

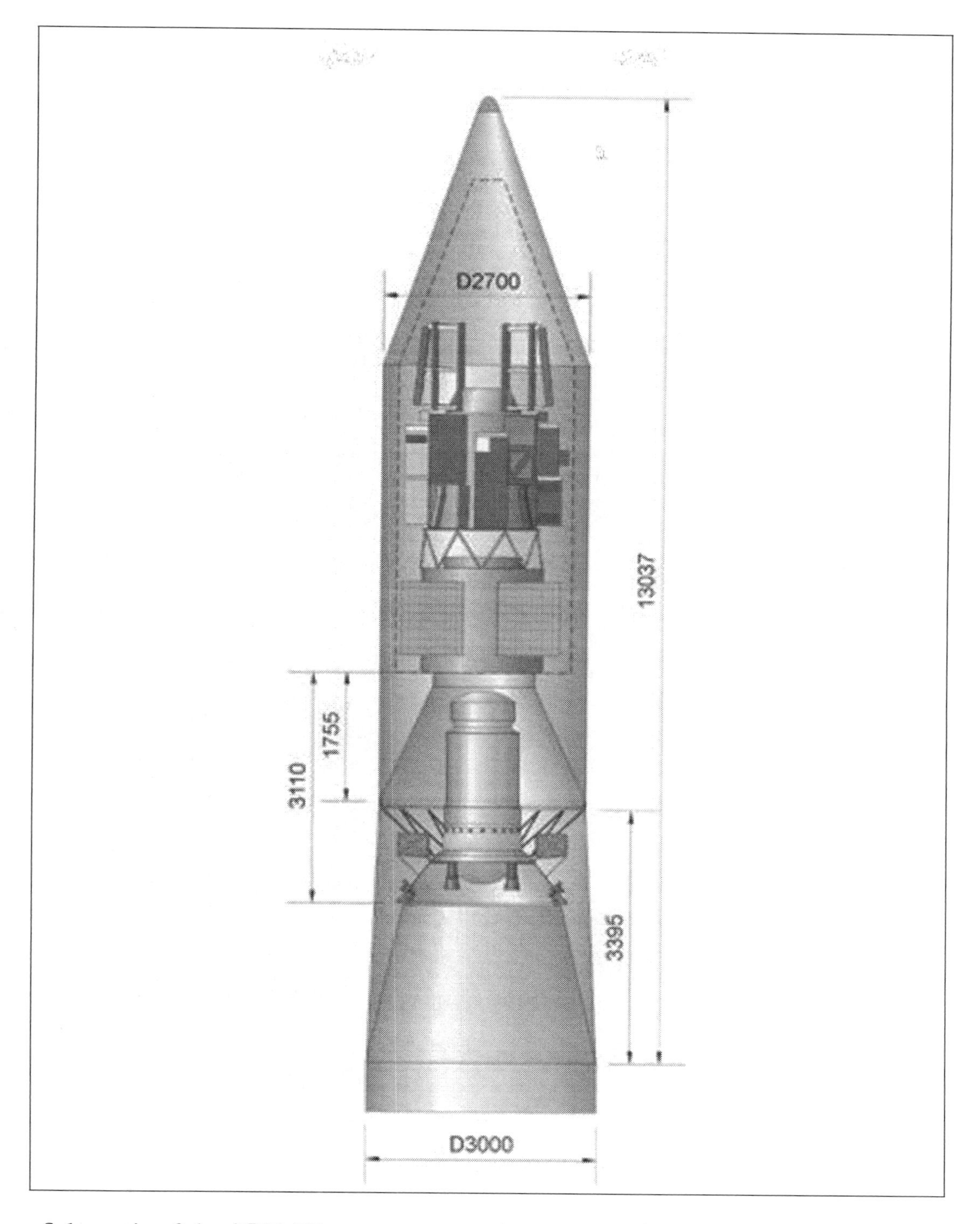

Schematic of the ADU-600 upper stage and nose cone of the Cyclone-2K showing a length of 13.037 m and an upper end diameter of 2.7 m. The Cyclone-2K was a further development of the Cyclone-2 for operations by the Russian Federation Space Agency, State Space Corporation ROSCOSMOS. State Space Corporation ROSCOSMOS

Diagram of the Cylclone-3 Launch vehicle (left) with an image of a Cyclone-2K during the launch sequence (right) for comparison. State Space Corporation ROSCOSMOS

The main payloads that the Cyclone-3 was intended to carry included Tselina, Meteor satellites and other spacecraft, 4000 kg able to be lifted to low and medium circular and elliptical Earth orbits. Plesetsk cosmodrome in Russia was designated as the launch site for the Cyclone-3; the first such launch taking place on 24 June 1977. The last of 122 launches was conducted on 30 January 2009, when the Kronos-Foton, payload was deployed. Overall the system proved to be reliable with 117 of the 122 launches being successful.

Cyclone-2 Launch Vehicle Specification – data furnished by Yuzhnoye State Design Office with input from the National Space Agency of Ukraine

Number of stages: 2
Length: 39000 mm
Stage diameter: 3000 mm
Main fairing height: 7.0 m
Main fairing diameter: 2.5 m
Payload fairing diameter: 2200 mm
Launch weight (for a 3.2 ton payload): 183 tons
Payload capability: 1.5-5 tons (National Space Agency of Ukraine states 3.2 tonnes maximum)
Propellants
Oxidizer: NTO
Fuel: UDMH
Vacuum thrust
Stage 1: 303.2 tnf
Stage 2: 101.5 tnf
Launch azimuth: 51-99°

Cyclone-3 Launch Vehicle Specification – data furnished by Yuzhnoye State Design Office with input from the National Space Agency of Ukraine

Number of stages: 3
Length: 39300 mm
Stage diameter: 3000 mm
Payload fairing height: 9.5 m
Payload fairing diameter: 2700 mm
Launch weight (for a 3.6 ton payload): 189 tons
Maximum payload mass: 3.6 tonnes to LEO or 0.6 tonnes to GTO
Propellants
Oxidizer: NTO
Fuel: UDMH
Thrust
Stage 1: 303.2 tnf
Stage 2: 101.5 tnf
Stage 3: 7.96 tnf
Payload capability to circular and elliptical orbits: Up to 4 tons
Launch azimuth: 73-83°

Top: Cyclone-2 launch vehicle. Above: Cyclone-3 launch vehicle. Yuzhnoye State Design Office

APPENDCIES

APPENDIX I

Soviet Orbital Weapon complex			
System	First Launch	Deployed	Retired
OKB-1 GR-1 8K713	NA	NA	NA
OKB-52 UR-200	4 November 1963	NA	NA
OBB-52 UR-500	16 July 1965	NA	NA
OKB-586 8K69	16 December 1965	1969	1983

APPENDIX II

Arms Treaty Range values for relative missile classes (excludes battlefield ballistic missiles)	
SRBM	Up to 599 nm (1109 km)
MRBM	600-1,499 nm (1111-2776 km)
IRBM	1,500-2,999 nm (2778-5554 km)
ICBM	in excess of 3,000 nm (5556 km)
DICBM	Intercontinental Class
FOBS	Global, in theory from any launch direction

APPENDIX III

Various designations for variants of the R-36 Missile Complex		
Manufacturer code	NATO/Treaty	Service
8K67	SS-9-1/SS-9-2 (SS-9 Mod1/Mod2)	R-36
8K67P	SS-9-4 (SS-9 Mod 4	R-36
8K69	SS-9-3/SS-X-6 (SS-9 Mod 3)	R-36 (R-36-O seldom used)

APPENDIX IV

Designations for the Cyclone-2 & Cyclone-3 launch vehicles		
Manufacturer code	Name	NATO Designation
11K69	Tsyclone-2/Cyclone-2	SS-X-6/SL-11
11K68	Tsyclone-3/Cyclone-3	SL-11

GLOSSARY

ABM	Anti-Ballistic Missile
Apoapsis	The farthest point reached by an orbiting object from the central body being orbited
Apogee	The farthest point reached by an orbiting object from the central body being orbited
ASAT	Anti-Satellite
Azimuth	The horizontal direction of a celestial point, such as an orbiting satellite, from a terrestrial point, the Earth
BMEWS	Ballistic Missile Early Warning System
CEP	Circular Error Probability
CONUS	Continental United States
Cosmos	Designation used for many Soviet and later Russian Federation space vehicles that operated in Earth orbit
DIA	Defense Intelligence Agency
DICBM	Depressed Trajectory Intercontinental Ballistic Missile
DoD	Department of Defense
Eccentricity	Measurement of the extent to which an orbiting object departs from an ellipse from a relative circle. This measurement is calculated as 1 full half of the distance of the two foci (the two fixed points of the ellipse) divided by the length of the semi-major axis. The calculated number is dimensionless, carrying values between 0 and 1.
EDB	Experimental Design Bureau
Epoch	Effectively the start of the timeline of a particular period of time or event
FOBS	Fractional Orbit Bombardment System
Gov.	Government
GR	Global Rocket
GTO	Geostationary Transfer Orbit – also known as a geosynchronous transfer orbit
HQ	Headquarters
ICBM	Intercontinental Ballistic Missile
ICS	Integrated Control System
II	Roman Numeral number 2
Inclination	In regards to an orbit, it refers to the angle created between the plane of the orbit and the ecliptic plane
INS	Inertial Navigation System
IOC	Initial Operational Capability
IRBM	Intermediate Range Ballistic Missile
kg	Kilogram
kgf	Kilogram force

kgf/s/kg	Kilogram force/per second/per kilogram
KIK	Ground Control Station
km	Kilometer
lb.	Pound – imperial weight
LEO	Low Earth Orbit
LRBM	Long Range Ballistic Missile
m	Meter
MAD	Mutually Assured Destruction
MIRV	Multiple Independently targetable Reentry Vehicle
mm	Millimeter
MOBS	Multiple Orbital Bombardment System
MODRF	Ministry of Defense of the Russian Federation
MRBM	Medium Range Ballistic Missile
MRV	Multiple Reentry Vehicle
MT	Megatons – 1 Megaton = 1,000.000 tons of TNT
NA	Not Applicable
NASA	National Aeronautics and Space Administration
NATO	North Atlantic Treaty Organization
NIE	National Intelligence Estimate
nm	Nautical Mile
NORAD	North American Air Defense
NRE	Non-Rotating Earth
NTO	Nitrogen Tetroxide
OGCh	Orbital'noy Govlnoy Chasti – Orbital Head Part
OKB	Design Bureau - Experimental Design Bureau
OKB-1	Experimental Design Bureau-1 (now S.P. Korolev Rocket and Space Corporation Energia)
OKB-52	Experimental Design Bureau-52 (now JSC MIC Mashinostryenia - Joint Stock Company Military Industrial Corporation Scientific and Production Machine Building Association
OKB-154	Experimental Design Bureau-154 (now OSC KBKhA - Konstruktorskoe Buro Khimavtomatiky)
OKB-456	Experimental Design Bureau-456 (now NPO Energomash)
OKB-586	Experimental Design Burea-586 (now Yuzhnoye State Design Office)
Periapsis	The closest approach point reached by an orbiting object from the central body being orbited
Period	The orbital period refers to the time taken for an object to complete one full revolution around its orbit. The time can be measured in seconds, minutes, days or years
R	Rocket
RV	Reentry Vehicle
SAC	Strategic Air Command

SMF	Strategic Missile Force
SRBM	Short Range Ballistic Missile
SS	NATO designation acronym for Soviet surface launched Ballistic Missiles
SSBN	Nuclear Powered Ballistic Missile Submarine
tnf	Ton force
UDMH	Unsymmetrical Dimethylhydrazine
US	United States
USSR	Union of Soviet Socialist Republics
UTC	Coordinated Universal Time
V	Vengeance
Warsaw Pact	A formal Treaty of Friendship, Co-operation and Mutual Assistance signed between the Socialist Republics of the USSR and 7 Soviet Orbit satellite states in Eastern Europe. This treaty, which took effect from 14 May 1955, was designed to counter the growing NATO alliance opposed to the Eastern Block
X	Experimental
±	Plus or minus
°	Degree

ABOUT THE AUTHOR

Hugh, a historian and author with extensive studies in scientific, aeronautic, astronautic and nautical technical and historical subjects, has published in excess of sixty books; non-fiction and fiction, writing under his given name as well as utilising two different pseudonyms. He has also written for several international magazines, whilst his work has been used as reference for many other projects ranging from the aviation industry, international news corporations and film media to encyclopaedias, museum exhibits and the computer gaming industry. He currently resides in his native Scotland

Other titles by the author include

Sukhoi T-50/PAK FA - Russia's 5th Generation 'Stealth' Fighter

Sukhoi Su-35S 'Flanker' E - Russia's 4++ Generation Super-Manoeuvrability Fighter

Sukhoi Su-34 'Fullback'

Sukhoi Su-30MKK/MK2/M2 - Russo Kitashiy Striker from Amur

MiG-35/D 'Fulcrum' F – Towards the Fifth Generation

Air War over Syria, Tu-160, Tu-95MS & Tu-22M3 - Cruise Missile and Bombing Strikes on Syria, November 2015-February 2016

Sukhoi Su-27SM(3)/SKM

Russian Non-Nuclear Attack Submarines – Project 877/877E/877EKM/Project 636/636.3 & Project 677/Amur 1650/950/S-1000

Russian/Soviet Aircraft Carrier & Carrier Aviation Design & Evolution Volume 1 - Seaplane Carriers, Project 71/72, Graf Zeppelin, Project 1123 ASW Cruiser & Project 1143-1143.4 Heavy Aircraft Carrying Cruiser

Light Battle Cruisers and the Second Battle of Heligoland Bight

British Battlecruisers of World War 1 - Operational Log, July 1914-June 1915

Eurofighter Typhoon - Storm over Europe

Tornado F.2/F.3 Air Defence Variant

Air to Air Missile Directory

North American F-108 Rapier - Mach 3 Interceptor

Convair YB-60 - Fort Worth Overcast

Boeing X-36 Tailless Agility Flight Research Aircraft

X-32 - The Boeing Joint Strike Fighter

X-35 - Progenitor to the F-35 Lightning II

X-45 Uninhabited Combat Air Vehicle

Into The Cauldron - The Lancaster MK.I Daylight Raid on Augsburg

Hurricane IIB Combat Log - 151 Wing RAF, North Russia 1941

RAF Meteor Jet Fighters in World War II, an Operational Log

Typhoon IA/B Combat Log - Operation Jubilee, August 1942

Defiant MK.I Combat Log - Fighter Command, May-September 1940

Blenheim MK.IF Combat Log - Fighter Command Day Fighter Sweeps/Night Interceptions, September 1939 - June 1940

Tomahawk I/II Combat Log - European Theatre, 1941-42

Fortress MK.I Combat Log - Bomber Command High Altitude Bombing Operations, July-September 1941

XF-92 - Convairs Arrow

Made in the USA
Middletown, DE
23 June 2017